AIAA AEROSPACE DESIGN ENGINEERS GUIDE

Fourth Edition

This book belongs to

_____

AIAA AEROSPACE DESIGN ENGINEERS GUIDE

Fourth Edition

American Institute of Aeronautics and Astronautics, Inc.
1801 Alexander Bell Drive
Reston, VA 20191-4344
September 1998

American Institute of Aeronautics and Astronautics, Inc., Reston, Virginia

Library of Congress Cataloging-in-Publication Data

AIAA aerospace design engineers guide.–4th ed.
    p. cm.
    1. Airplanes—Design and construction—Handbooks, manuals, etc.
  2. Space vehicles—Design and construction—Handbooks, manuals, etc.
 I. American Institute of Aeronautics and Astronautics.
  TL551.A32   1998   629.1'2—dc21   98-34952
                                                       CIP

    ISBN 1-56347-283-X (softcover : alk. paper)

Foreword

This fourth edition of the *AIAA Aerospace Design Engineers Guide* (ADEG) is a revised and enlarged version of the previous edition. It has been prepared under the charter of the AIAA Design Engineering Technical Committee to assist the design engineer in creating and defining practical aerospace products. The intended scope of this guide is to provide a condensed collection of commonly used engineering reference data and to also function as a general reference guide for disciplines related specifically to aerospace design.

The previous editions were published in 1983, 1987, and 1993, and the guide will be updated whenever the committee has accumulated sufficient material to warrant a new edition. The physical size of the fourth edition has been enlarged and rearranged to enhance user access and utilization.

Materials included in the guide were compiled principally from design manuals and handbooks. The Design Engineering Technical Committee is indebted to many people and companies for their voluntary cooperation.

The committee does not guarantee the accuracy of the information in this guide, and it should not be referenced as a final authority for certification of designs. We solicit your comments and suggestions for improvement, including corrections and candidate new materials, so that future editions can be made more useful for the needs of the design engineering community.

ADEG Subcommittee
AIAA Design Engineering Technical Committee
1801 Alexander Bell Drive
Reston, VA 20191-4344

Table of Contents

Preface

Using This Guide

The *AIAA Aerospace Design Engineers Guide* has been compiled to assist aerospace design engineers during design inception and development. The purpose of this guide is to serve as a general purpose, in-field handbook used by design engineers to perform back-of-the-envelope/rough-order-of-magnitude estimates and calculations for early preliminary and conceptual aerospace design. The guide is not intended to be a comprehensive handbook for producing highly detailed production designs, although some of the design data may be suitable for the designer's objectives. Other specialized handbooks and detailed company handbooks/manuals and specifications are generally available to the designer to complete comprehensive detail and production design efforts.

This guide is divided into 10 major sections, as shown in the Table of Contents, each with a summary topic list. The sections provide the following categories of aerospace design engineering information. Sections 1, 2, and 3 provide mathematical definition, section properties, and conversion factors. Sections 4, 5, and 6 provide detail information on structures, mechanical, and electrical design. Sections 8 and 9 provide universal product definition nomenclature in the form of "geometric dimensioning and tolerancing" and general materials and specifications. Sections 7 and 10 provide specific aerospace design for aircraft and spacecraft.

A list of topics are included at the beginning of each section to help the user quickly and easily find information. The 10 sections of the Design Guide are arranged to maximize design data, which appear in the form of visual and written explanations, tabular data, formulas/equations, graphics, figures, maps, glossaries, standards, specifications, references, and design rules of thumb.

Introduction

Design engineering—a discipline that creates and transforms ideas and concepts into a product definition that satisfies customer requirements

Perceptions of design engineering vary greatly in the aerospace industry, varying from design drafting to coordinated control over the product, from the conceptual phase to the finished product. The introduction of concurrent engineering (CE), or integrated product development (IPD), emphasizes the importance of design engineers as a focal point for a myriad of inputs from the engineering, development, and production areas to produce a high-quality, cost-effective, properly functioning, competitive product.

Design engineering is fundamental to every project. It is a discipline that creates and transforms ideas into a product definition that satisfies customer requirements. The role of the design engineer is the creation, synthesis, iteration, and presentation of design solutions. The design engineer coordinates with engineering specialists and integrates their inputs to produce the form, fit, and function documentation to completely define the product.

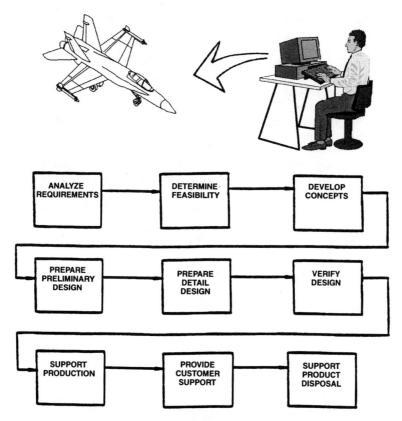

Design engineering is a multidisciplined iterative process.

Simply stated, in the integrated product development process, the design engineer interacts with the customer, the project manager, the systems integration personnel, and the production team. Interactions and communications result in the integration of program performance requirements, specialty engineering inputs, and, of course, cost and schedule impacts. All of these inputs affect conceptual tradeoffs and final depiction of the design of the product.

The design engineer must be able to identify the critical requirements from the multiple, even conflicting, constraints of the initial request to create appropriate functional concepts. The design engineer must conceive of many different ways to fulfill the requirements. Several conceptual solutions are required to evaluate and select the best one. The ability to synthesize design concepts and present and explain them to others is a crucial skill for design engineers.

The design engineer integrates the many specialty disciplines of the engineering domain that are required for the evaluation and selection of the options and design trades that are part of the new product performance iteration process. Satisfactory products are developed when the design engineer successfully integrates, by analysis and documentation, the feasible capabilities so that all requirements are met, including manufacturability, quality assurance, product safety, reliability, maintainability, and human engineering. Schedule and cost goals must also be met for both the design process and the resulting product. When used, the product, and design, will be judged and each must perform on its own merits.

The ultimate responsibility for the delivery of a well-designed and properly functioning product rests with the design engineer. Design engineering seeks an optimal whole, rather than attempting to perfect each individual part within a system, thus obtaining a balanced, designed product that fulfills the requirements and satisfies the customer.

The design engineering process is an iterative one. Iteration is necessary during each phase of the design process to coordinate and improve previous decisions, designs, and functions. Ideally, these iterations must be kept to a minimum—hence, the IPD process. However, some modifications after major reviews or verification testing during the program are not unusual and are coordinated through the design engineer, from analyzing requirements and determining feasibility, all the way through supporting the product disposal when that event becomes necessary. It is easy to see that iterations in the design could occur in many of the design tasks shown.

In essence, design engineering is the creative process in which ideas of one or more contributors are converted from ideas to documents that define a product that can be profitably manufactured and that meet the design performance and functioning specifications.

Source: AIAA Design Engineering Technical Committee

Section 1

MATHEMATICS

Greek Alphabet

Capital	Lowercase	Letter	Capital	Lowercase	Letter
A	α	Alpha	N	ν	Nu
B	β	Beta	Ξ	ξ	Xi
Γ	γ	Gamma	O	o	Omicron
Δ	δ	Delta	Π	π	Pi
E	ϵ	Epsilon	P	ρ	Rho
Z	ζ	Zeta	Σ	σ	Sigma
H	η	Eta	T	τ	Tau
Θ	θ	Theta	Υ	υ	Upsilon
I	ι	Iota	Φ	ϕ	Phi
K	κ	Kappa	X	χ	Chi
Λ	λ	Lambda	Ψ	ψ	Psi
M	μ	Mu	Ω	ω	Omega

SI Prefixes

Multiplication factor	Prefix	Symbol
$1\ 000\ 000\ 000\ 000\ 000\ 000\ 000\ 000 = 10^{24}$	yotta	Y
$1\ 000\ 000\ 000\ 000\ 000\ 000\ 000\ 000 = 10^{21}$	zetta	Z
$1\ 000\ 000\ 000\ 000\ 000\ 000 = 10^{18}$	exa	E
$1\ 000\ 000\ 000\ 000\ 000 = 10^{15}$	peta	P
$1\ 000\ 000\ 000\ 000 = 10^{12}$	tera	T
$1\ 000\ 000\ 000 = 10^{9}$	giga	G
$1\ 000\ 000 = 10^{6}$	mega	M
$1\ 000 = 10^{3}$	kilo	k
$100 = 10^{2}$	hecto[a]	h
$10 = 10^{1}$	deka[a]	da
$0.1 = 10^{-1}$	deci[a]	d
$0.01 = 10^{-2}$	centi[a]	c
$0.001 = 10^{-3}$	milli	m
$0.000\ 001 = 10^{-6}$	micro	μ
$0.000\ 000\ 001 = 10^{-9}$	nano	n
$0.000\ 000\ 000\ 001 = 10^{-12}$	pico	p
$0.000\ 000\ 000\ 000\ 001 = 10^{-15}$	femto	f
$0.000\ 000\ 000\ 000\ 000\ 001 = 10^{-18}$	atto	a
$0.000\ 000\ 000\ 000\ 000\ 000\ 001 = 10^{-21}$	zepto	z
$0.000\ 000\ 000\ 000\ 000\ 000\ 000\ 001 = 10^{-24}$	yocto	y

[a]To be avoided where possible.

Algebra

Powers and Roots

$$a^n = a \cdot a \cdot a \ldots \text{to } n \text{ factors} \qquad a^{-n} = \frac{1}{a^n}$$

$$a^m \cdot a^n = a^{m+n} \qquad \frac{a^m}{a^n} = a^{m-n}$$

$$(ab)^n = a^n b^n \qquad \left(\frac{a}{b}\right)^n = \frac{a^n}{b^n}$$

$$(a^m)^n = (a^n)^m = a^{mn} \qquad (\sqrt[n]{a})^n = a$$

$$a^{1/n} = \sqrt[n]{a} \qquad a^{m/n} = \sqrt[n]{a^m}$$

$$\sqrt[n]{ab} = \sqrt[n]{a}\sqrt[n]{b} \qquad \sqrt[n]{\frac{a}{b}} = \frac{\sqrt[n]{a}}{\sqrt[n]{b}}$$

$$\sqrt[n]{\sqrt[m]{a}} = \sqrt[mn]{a}$$

Zero and Infinity Operations

$a \cdot 0 = 0$	$a \cdot \infty = \infty$	$0 \cdot \infty$	indeterminate
$\dfrac{0}{a} = 0$	$\dfrac{a}{0} = \infty$	$\dfrac{0}{0}$	indeterminate
$\dfrac{\infty}{a} = \infty$	$\dfrac{a}{\infty} = 0$	$\dfrac{\infty}{\infty}$	indeterminate
$a^0 = 1$	$0^a = 0$	0^0	indeterminate
$\infty^a = \infty$		∞^0	indeterminate
$a - a = 0$	$\infty - a = \infty$	$\infty - \infty$	indeterminate

$$a^\infty = \infty \quad \text{if } a^2 > 1 \qquad a^\infty = 0 \quad \text{if } a^2 < 1 \qquad a^\infty = 1 \quad \text{if } a^2 = 1$$
$$a^{-\infty} = 0 \quad \text{if } a^2 > 1 \qquad a^{-\infty} = \infty \quad \text{if } a^2 < 1 \qquad a^{-\infty} = 1 \quad \text{if } a^2 = 1$$

Binomial Expansions

$$(a \pm b)^2 = a^2 \pm 2ab + b^2$$

$$(a \pm b)^3 = a^3 \pm 3a^2 b + 3ab^2 \pm b^3$$

$$(a \pm b)^4 = a^4 \pm 4a^3 b + 6a^2 b^2 \pm 4ab^3 + b^4$$

$$(a \pm b)^n = a^n \pm \frac{n}{1} a^{n-1} b + \frac{n(n-1)}{1 \cdot 2} a^{n-2} b^2$$

$$\pm \frac{n(n-1)(n-2)}{1 \cdot 2 \cdot 3} a^{n-3} b^3 + \cdots$$

Note: n may be positive or negative, integral or fractional. If n is a positive integer, the series has $(n+1)$ terms; otherwise, the number of terms is infinite.

Algebra, continued

Logarithms

$$\log_b b = 1, \ \log_b 1 = 0, \ \log_b 0 = \begin{cases} +\infty & \text{when } b \text{ lies between 0 and 1} \\ -\infty & \text{when } b \text{ lies between 1 and } \infty \end{cases}$$

$$\log_b M \cdot N = \log_b M + \log_b N \qquad \log_b \frac{M}{N} = \log_b M - \log_b N$$

$$\log_b N^p = p \log_b N \qquad \log_b \sqrt{N^p} = \frac{p}{r} \log_b N$$

$$\log_b N = \frac{\log_a N}{\log_a b} \qquad \log_b b^N = N \quad b^{\log_b N} = N$$

The Quadratic Equation

If

$$ax^2 + bx + c = 0$$

then

$$x = \frac{-b \pm \sqrt{b^2 - 4ac}}{2a} = \frac{2c}{-b \mp \sqrt{b^2 - 4ac}}$$

The second equation serves best when the two values of x are nearly equal.

$$\text{If } b^2 - 4ac = 0 \ \begin{matrix} > \\ \\ < \end{matrix} \begin{cases} \text{the roots are real and unequal} \\ \text{the roots are real and equal} \\ \text{the roots are imaginary} \end{cases}$$

The Cubic Equations

Any cubic equation $y^3 + py^2 + qy + r = 0$ may be reduced to the form $x^3 + ax + b = 0$ by substituting for y the value $[x - (p/3)]$. Here $a = 1/3(3q - p^2)$, $b = 1/27(2p^3 - 9pq + 27r)$.

Algebraic Solution of $x^3 + ax + b = 0$

$$A = \sqrt[3]{-\frac{b}{2} + \sqrt{\frac{b^2}{4} + \frac{a^3}{27}}} \qquad B = \sqrt[3]{-\frac{b}{2} - \sqrt{\frac{b^2}{4} + \frac{a^3}{27}}}$$

$$x = A + B \qquad -\frac{A+B}{2} + \frac{A-B}{2}\sqrt{-3} \qquad -\frac{A+B}{2} - \frac{A-B}{2}\sqrt{-3}$$

If

$$\frac{b^2}{4} + \frac{a^3}{27} = 0 \ \begin{matrix} > \\ \\ < \end{matrix} \begin{cases} \text{1 real root, 2 conjugate imaginary roots} \\ \text{3 real roots, at least 2 equal} \\ \text{3 real and unequal roots} \end{cases}$$

Algebra, continued

Trigonometric Solution for x³ + ax + b = 0

Where $(b^2/4) + (a^3/27) < 0$, these formulas give the roots in impractical form for numerical computation. (In this case, a is negative.) Compute the value of angle ϕ derived from

$$\cos\phi = \sqrt{\frac{b^2}{4} \div \left(-\frac{a^3}{27}\right)}$$

Then

$$x = \mp 2\sqrt{-\frac{a}{3}}\cos\frac{\phi}{3} \qquad \mp 2\sqrt{-\frac{a}{3}}\cos\left(\frac{\phi}{3} + 120°\right)$$

$$\mp 2\sqrt{-\frac{a}{3}}\cos\left(\frac{\phi}{3} + 240°\right)$$

where the upper or lower signs describe b as positive or negative.

Where $(b^2/4) + (a^3/27) >= 0$, compute the values of the angles ψ and ϕ from $\cot 2\psi = [(b^2/4) \div (a^3/27)]^{1/2}$, $\tan\phi = (\tan\psi)^{1/3}$. The real root of the equation then becomes

$$x = \pm 2\sqrt{\frac{a}{3}}\cot 2\phi$$

where the upper or lower sign describes b as positive or negative.

When $(b^2/4) + (a^3/27) = 0$, the roots become

$$x = \mp 2\sqrt{-\frac{a}{3}} \qquad \pm\sqrt{-\frac{a}{3}} \qquad \pm\sqrt{-\frac{a}{3}}$$

where the upper or lower signs describe b as positive or negative.

The Binomial Equation

When $x^n = a$, the n roots of this equation become
   a positive:

$$x = \sqrt[n]{a}\left(\cos\frac{2k\pi}{n} + \sqrt{-1}\sin\frac{2k\pi}{n}\right)$$

   a negative:

$$x = \sqrt[n]{-a}\left[\cos\frac{(2k+1)\pi}{n} + \sqrt{-1}\sin\frac{(2k+1)\pi}{n}\right]$$

where k takes in succession values of $0, 1, 2, 3, \ldots, n - 1$.

Trigonometry

Trigonometric Functions of an Angle

$$\text{sine (sin)}\,\alpha = \frac{y}{r}$$

$$\text{cosine (cos)}\,\alpha = \frac{x}{r}$$

$$\text{tangent (tan)}\,\alpha = \frac{y}{x}$$

$$\text{cotangent (cot)}\,\alpha = \frac{x}{y}$$

$$\text{secant (sec)}\,\alpha = \frac{r}{x}$$

$$\text{cosecant (csc)}\,\alpha = \frac{r}{y}$$

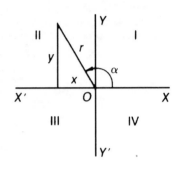

Signs of Functions

Quadrant	sin	cos	tan	cot	sec	csc
I	+	+	+	+	+	+
II	+	−	−	−	−	+
III	−	−	+	+	−	−
IV	−	+	−	−	+	−

Functions of 0°, 30°, 45°, 60°, 90°, 180°, 270°, 360°

	0°	30°	45°	60°	90°	180°	270°	360°
sin	0	$1/2$	$\sqrt{2}/2$	$\sqrt{3}/2$	1	0	−1	0
cos	1	$\sqrt{3}/2$	$\sqrt{2}/2$	$1/2$	0	−1	0	1
tan	0	$\sqrt{3}/3$	1	$\sqrt{3}$	∞	0	∞	0
cot	∞	$\sqrt{3}$	1	$\sqrt{3}/3$	0	∞	0	∞
sec	1	$2\sqrt{3}/3$	$\sqrt{2}$	2	∞	−1	∞	1
csc	∞	2	$\sqrt{2}$	$2\sqrt{3}/3$	1	∞	−1	∞

Fundamental Function Relations

$\sin\alpha = 1/\csc\alpha$ $\cos\alpha = 1/\sec\alpha$ $\tan = 1/\cot\alpha = \sin\alpha/\cos\alpha$

$\csc\alpha = 1/\sin\alpha$ $\sec\alpha = 1/\cos\alpha$ $\cot\alpha = 1/\tan\alpha = \cos\alpha/\sin\alpha$

$\sin^2\alpha + \cos^2\alpha = 1$ $\sec^2\alpha - \tan^2\alpha = 1$ $\csc^2\alpha - \cot^2\alpha = 1$

Trigonometry, continued

Functions of Several Angles

$$\sin 2\alpha = 2\sin\alpha\cos\alpha$$

$$\cos 2\alpha = 2\cos^2\alpha - 1 = 1 - 2\sin^2\alpha = \cos^2\alpha - \sin^2\alpha$$

$$\sin 3\alpha = 3\sin\alpha - 4\sin^3\alpha$$

$$\cos 3\alpha = 4\cos^3\alpha - 3\cos\alpha$$

$$\sin 4\alpha = 4\sin\alpha\cos\alpha - 8\sin^3\alpha\cos\alpha$$

$$\cos 4\alpha = 8\cos^4\alpha - 8\cos^2\alpha + 1$$

$$\sin n\alpha = 2\sin(n-1)\alpha\cos\alpha - \sin(n-2)\alpha$$

$$\cos n\alpha = 2\cos(n-1)\alpha\cos\alpha - \cos(n-2)\alpha$$

Half-Angle Functions

$$\sin\frac{\alpha}{2} = \sqrt{\frac{1-\cos\alpha}{2}} \qquad \cos\frac{1}{2}\alpha = \sqrt{\frac{1+\cos\alpha}{2}}$$

$$\tan\frac{1}{2}\alpha = \frac{1-\cos\alpha}{\sin\alpha} = \frac{\sin\alpha}{1+\cos\alpha} = \sqrt{\frac{-\cos\alpha}{1+\cos\alpha}}$$

Powers of Functions

$$\sin^2\alpha = \tfrac{1}{2}(1-\cos 2\alpha) \qquad\qquad \cos^2\alpha = \tfrac{1}{2}(1+\cos 2\alpha)$$

$$\sin^3\alpha = \tfrac{1}{4}(3\sin\alpha - \sin 3\alpha) \qquad\qquad \cos^3\alpha = \tfrac{1}{4}(\cos 3\alpha + 3\cos\alpha)$$

$$\sin^4\alpha = \tfrac{1}{8}(\cos 4\alpha - 4\cos 2\alpha + 3) \qquad \cos^4\alpha = \tfrac{1}{8}(\cos 4\alpha + 4\cos 2\alpha + 3)$$

Functions: Sum or Difference of Two Angles

$$\sin(\alpha \pm \beta) = \sin\alpha\cos\beta \pm \cos\alpha\sin\beta$$

$$\cos(\alpha \pm \beta) = \cos\alpha\cos\beta \mp \sin\alpha\sin\beta$$

$$\tan(\alpha \pm \beta) = \frac{\tan\alpha \pm \tan\beta}{1 \mp \tan\alpha\tan\beta}$$

Trigonometry, continued

Sums, Differences, and Products of Two Functions

$$\sin \alpha \pm \sin \beta = 2 \sin \tfrac{1}{2}(\alpha \pm \beta)\cos \tfrac{1}{2}(\alpha \mp \beta)$$

$$\cos \alpha + \cos \beta = 2 \cos \tfrac{1}{2}(\alpha + \beta)\cos \tfrac{1}{2}(\alpha - \beta)$$

$$\cos \alpha - \cos \beta = -2 \sin \tfrac{1}{2}(\alpha + \beta)\sin \tfrac{1}{2}(\alpha - \beta)$$

$$\tan \alpha \pm \tan \beta = \frac{\sin(\alpha \pm \beta)}{\cos \alpha \cos \beta}$$

$$\sin^2 \alpha - \sin^2 \beta = \sin(\alpha + \beta)\sin(\alpha - \beta)$$

$$\cos^2 \alpha - \cos^2 \beta = -\sin(\alpha + \beta)\sin(\alpha - \beta)$$

$$\cos^2 \alpha - \sin^2 \beta = \cos(\alpha + \beta)\cos(\alpha - \beta)$$

$$\sin \alpha \sin \beta = \tfrac{1}{2}\cos(\alpha - \beta) - \tfrac{1}{2}\cos(\alpha + \beta)$$

$$\cos \alpha \cos \beta = \tfrac{1}{2}\cos(\alpha - \beta) + \tfrac{1}{2}\cos(\alpha + \beta)$$

$$\sin \alpha \cos \beta = \tfrac{1}{2}\sin(\alpha + \beta) + \tfrac{1}{2}\sin(\alpha - \beta)$$

Right Triangle Solution

Given any two sides, or one side and any acute angle α, find the remaining parts:

$$\sin \alpha = \frac{a}{c} \qquad \cos \alpha = \frac{b}{c} \qquad \tan \alpha = \frac{a}{b} \qquad \beta = 90° - \alpha$$

$$a = \sqrt{(c + b)(c - b)} = c \sin \alpha = b \tan \alpha$$

$$b = \sqrt{(c + a)(c - a)} = c \cos \alpha = \frac{a}{\tan \alpha}$$

$$c = \frac{a}{\sin \alpha} = \frac{b}{\cos \alpha} = \sqrt{a^2 + b^2}$$

$$A = \frac{1}{2}ab = \frac{a^2}{2 \tan \alpha} = \frac{b^2 \tan \alpha}{2} = \frac{c^2 \sin 2\alpha}{4}$$

Mensuration

Note: A = area, V = volume

Oblique Triangle Solution
Right Triangle

$$A = \tfrac{1}{2} ab \qquad c = \sqrt{a^2 + b^2}$$
$$a = \sqrt{c^2 - b^2} \qquad b = \sqrt{c^2 - a^2}$$

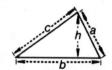

Oblique Triangle

$$A = \tfrac{1}{2} bh$$

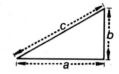

Equilateral Triangle

$$A = \tfrac{1}{2} ah = \tfrac{1}{4} a^2 \sqrt{3}$$
$$h = \tfrac{1}{2} a \sqrt{3}$$
$$r_1 = \frac{a}{2\sqrt{3}}$$
$$r_2 = \frac{a}{\sqrt{3}}$$

Parallelogram

$$A = ah = ab \sin \alpha$$
$$d_1 = \sqrt{a^2 + b^2 - 2ab \cos \alpha}$$
$$d_2 = \sqrt{a^2 + b^2 + 2ab \cos \alpha}$$

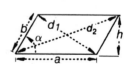

Trapezoid

$$A = \tfrac{1}{2} h(a + b)$$

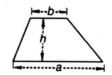

Mensuration, continued

Circle

C = circumference

α = central angle in radians

$C = \pi D = 2\pi R$

$$c = R\alpha = \frac{1}{2}D\alpha = D\cos^{-1}\frac{d}{R} = D\tan^{-1}\frac{l}{2d}$$

$$l = 2\sqrt{R^2 - d^2} = 2R\sin\frac{\alpha}{2} = 2d\tan\frac{\alpha}{2} = 2d\tan\frac{c}{D}$$

$$d = \frac{1}{2}\sqrt{4R^2 - l^2} = \frac{1}{2}\sqrt{D^2 - l^2} = R\cos\frac{\alpha}{2}$$

$$= \frac{1}{2}l\cot\frac{\alpha}{2} = \frac{1}{2}\cot\frac{c}{D}$$

$$h = R - d$$

$$\alpha = \frac{c}{R} = \frac{2c}{D} = 2\cos^{-1}\frac{d}{R}$$

$$= 2\tan^{-1}\frac{l}{2d} = 2\sin^{-1}\frac{l}{D}$$

$$A = \pi R^2 = \tfrac{1}{4}\pi D^2 = \tfrac{1}{2}RC = \tfrac{1}{4}DC$$

$$A_{\text{sector}} = \tfrac{1}{2}Rc = \tfrac{1}{2}R^2\alpha = \tfrac{1}{8}D^2\alpha$$

$$A_{\text{segment}} = A_{\text{sector}} - A_{\text{triangle}} = \frac{1}{2}R^2(\alpha - \sin\alpha)$$

$$= \frac{1}{2}R\left(c - R\sin\frac{c}{R}\right)$$

$$= R^2\sin^{-1}\frac{l}{2R} - \frac{1}{4}l\sqrt{4R^2 - l^2}$$

$$= R^2\cos^{-1}\frac{d}{R} - d\sqrt{R^2 - d^2}$$

$$= R^2\cos^{-1}\frac{R-h}{R} - (R-h)\sqrt{2Rh - h^2}$$

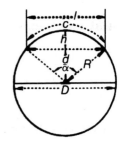

Mensuration, continued

Area by Approximation

If $y_0, y_1, y_2, \ldots, y_n$ are the lengths of a series of equally spaced parallel chords and if h is their distance apart, the area enclosed by boundary is given approximately by

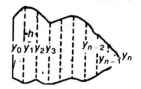

$$A_s = \tfrac{1}{3}h[(y_0 + y_n) + 4(y_1 + y_3 + \cdots + y_{n-1}) + 2(y_2 + y_4 + \cdots + y_{n-1})]$$

where n is even (Simpson's Rule).

Ellipse

$$A = \pi ab/4$$

Perimeter s

$$= \pi(a+b)\left[1 + \frac{1}{4}\left(\frac{a-b}{a+b}\right)^2 + \frac{1}{64}\left(\frac{a-b}{a+b}\right)^4 + \frac{1}{256}\left(\frac{a-b}{a+b}\right)^6 + \cdots\right]$$

$$\approx \pi\frac{a+b}{4}\left[3(1+\lambda) + \frac{1}{1-\lambda}\right]$$

where $\lambda = \left[\dfrac{a-b}{2(a+b)}\right]^2$

Parabola

$$A = (2/3)ld$$

Length of arc s

$$= \frac{1}{2}\sqrt{16d^2 + l^2} + \frac{l^2}{8d}\ell_n\left(\frac{4d + \sqrt{16d^2 + l^2}}{l}\right)$$

$$= l\left[1 + \frac{2}{3}\left(\frac{2d}{l}\right)^2 - \frac{2}{5}\left(\frac{2d}{l}\right)^4 + \cdots\right]$$

Height of segment $d_1 = \dfrac{d}{l^2}(l^2 - l_1^2)$

Width of segment $l_1 = l\sqrt{\dfrac{d - d_1}{d}}$

MATHEMATICS

Mensuration, continued

Catenary

Length of arc s

$$= 1\left[1 + \frac{2}{3}\left(\frac{2d}{l}\right)^2\right]$$

approximately, if d is small in comparison with l.

Sphere

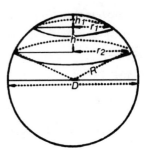

$$A = 4\pi R^2 = \pi D^2$$

$$A_{\text{zone}} = 2\pi Rh = \pi Dh$$

$$V_{\text{sphere}} = 4/3\pi R^3 = 1/6\pi D^3$$

$$V_{\text{spherical sector}} = 72/3\pi R^2h = 1/6\pi D^2h$$

V for spherical segment of one base:

$$1/6\pi h\left(3r_1^2 + 3r_2^2 + h^2\right)$$

Ellipsoid

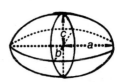

$$V = 4/3\pi abc$$

Paraboloid Segment

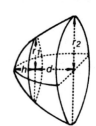

$$V(\text{segment of one base}) = \tfrac{1}{2}\pi r_1^2 h$$

$$V(\text{segment of two bases}) = \tfrac{1}{2}\pi d\left(r_1^2 + r_2^2\right)$$

Torus

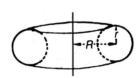

$$V = 2\pi^2 Rr^2$$

$$\text{Surface} = 4\pi^2 Rr$$

Analytic Geometry

Rectangular and Polar Coordinate Relations

$$x = r \cos \theta \qquad y = r \sin \theta$$

$$r = \sqrt{x^2 + y^2} \qquad \theta = \tan^{-1} \frac{y}{x} \qquad \sin \theta = \frac{y}{\sqrt{x^2 + y^2}}$$

$$\cos \theta = \frac{x}{\sqrt{x^2 + y^2}} \qquad \tan \theta = \frac{y}{x}$$

Points and Slopes

For $P_1(x_1, y_1)$ and $P_2(x_2, y_2)$ as any two points and α the angle from OX to $P_1 P_2$ measured counterclockwise,

$$P_1 P_2 = d = \sqrt{(x_2 - x_1)^2 + (y_2 - y_1)^2}$$

$$P_1 P_2 \text{ midpoint} = \frac{x_1 + x_2}{2}, \frac{y_1 + y_2}{2}$$

For point m_2 that divides $P_1 P_2$ in the ratio m_1,

$$\left(\frac{m_1 x_2 + m_2 x_1}{m_1 + m_2}, \frac{m_1 y_2 + m_2 y_1}{m_1 + m_2} \right)$$

$$\text{Slope of } P_1 P_2 = \tan \alpha = m = \frac{y_2 - y_1}{x_2 - x_1}$$

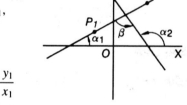

For angle β between two lines of slopes m_1 and m_2,

$$\beta = \tan^{-1} \frac{m_2 - m_1}{1 + m_1 m_2}$$

Straight Line

$$Ax + By + C = 0$$

$$-A \div B = \text{slope}$$

$$y = mx + b$$

(where $m = $ slope, $b = $ intercept on OY)

$$y - y_1 = m(x - x_1)$$

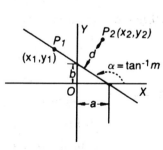

[$m = $ slope, where $P_1(x_1, y_1)$ is a known point on line]

$$d = \frac{Ax_2 + By_2 + C}{\pm \sqrt{A^2 + B^2}}$$

[where d equals distance from a point $P_2(x_2, y_2)$ to the line $Ax + By + C = 0$]

Analytic Geometry, continued

Circle

Locus of a point at a constant distance (radius) from a fixed point C (center).

Circle A

$$(x - h)^2 + (y - k)^2 = a^2$$

$$r^2 + b^2 - 2br \cos(\theta - \beta) = a^2$$

$C(h, k),\ \text{rad} = a \qquad C(b, \beta),\ \text{rad} = a$

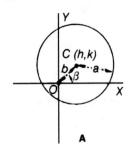

A

Circle B$_{bottom}$

$$x^2 + y^2 = 2ax$$

$$r = 2a \cos \theta$$

$C(a, 0),\ \text{rad} = a \qquad C(a, 0),\ \text{rad} = a$

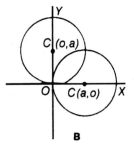

Circle B$_{top}$

$$x^2 + y^2 = 2ay$$

$$r = 2a \sin \theta$$

$C(0, a),\ \text{rad} = a \qquad C\left(a, \dfrac{\pi}{2}\right),\ \text{rad} = a$

B

Circle C

$$x^2 + y^2 = a^2$$

$$r = a$$

$x = a \cos \phi \qquad y = a \sin \phi$

$C(0, 0),\ \text{rad} = a$

$C(0, 0),\ \text{rad} = a$

$C(0, 0),\ \text{rad} = a \qquad \phi = \text{angle from } OX \text{ to radius}$

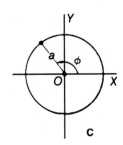

C

Analytic Geometry, continued

Parabola

Conic where $e = 1$; lastus rectum, a.

Parabola A

$(y - k)^2 = a(x - h)$ Vertex (h, k), axis $\parallel OX$

$y^2 = ax$ Vertex $(0, 0)$, axis along OX

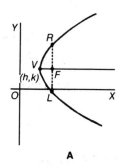

A

Parabola B

$(x - h)^2 = a(y - k)$ Vertex (h, k), axis $\parallel OY$

$x^2 = ay$ Vertex $(0, 0)$, axis along OY

Distance from vertex to focus: $\frac{1}{4} a$

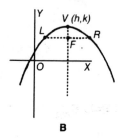

B

Sine Wave

$y = a \sin(bx + c)$

$y = a \cos(bx + c') = a \sin(bx + c)$ (where $c = c' + \pi/2$)

$y = m \sin bx + n \cos bx = a \sin(bx + c)$

     (where $a = \sqrt{m^2 + n^2}, c = \tan^{-1} n/m$)

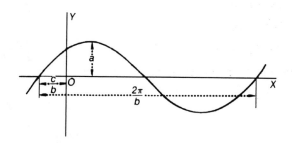

Analytic Geometry, continued

Exponential or Logarithmic Curves

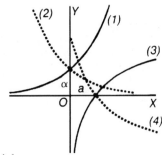

1) $y = ab^x$ or $x = \log_b \dfrac{y}{a}$

2) $y = ab^{-x}$ or $x = -\log_b \dfrac{y}{a}$

3) $x = ab^y$ or $y = \log_b \dfrac{x}{a}$

4) $x = ab^{-y}$ or $y = -\log_b \dfrac{x}{a}$

The equations $y = ae^{\pm nx}$ and $x = ae^{\pm ny}$ are special cases.

Catenary

Line of uniform weight suspended freely between two points at the same level:

$$y = a\left[\cosh\left(\frac{x}{a}\right) - 1\right] = \frac{a}{2}\left(e^{x/a} + e^{-x/a} - 2a\right)$$

(where y is measured from the lowest point of the line, w is the weight per unit length of the line, T_h is the horizontal component of the tension in the line, and a is the parameter of the catenary; $a = T_h/w$)

Helix

Curve generated by a point moving on a cylinder with the distance it transverses parallel to the axis of the cylinder being proportional to the angle of rotation about the axis:

$$x = a\cos\theta$$
$$y = a\sin\theta$$
$$z = k\theta$$

(where $a =$ radius of cylinder, $2\pi k =$ pitch)

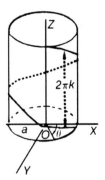

Analytic Geometry, continued

Points, Lines, and Planes

Distance between points $P_1(x_1, y_1, z_1)$ and $P_2(x_2, y_2, z_2)$:

$$d = \sqrt{(x_2 - x_1)^2 + (y_2 - y_1)^2 + (z_2 - z_1)^2}$$

Direction cosine of a line (cosines of the angles α, β, γ which the line or any parallel line makes with the coordinate axes) is related by

$$\cos^2 \alpha + \cos^2 \beta + \cos^2 \gamma = 1$$

If $\cos \alpha : \cos \beta : \cos \gamma = a : b : c$, then

$$\cos \alpha = \frac{a}{\sqrt{a^2 + b^2 + c^2}} \qquad \cos \beta = \frac{b}{\sqrt{a^2 + b^2 + c^2}}$$

$$\cos \gamma = \frac{c}{\sqrt{a^2 + b^2 + c^2}}$$

Direction cosine of the line joining $P_1(x_1, y_1, z_1)$ and $P_2(x_2, y_2, z_2)$:

$$\cos \alpha : \cos \beta : \cos \gamma = x_2 - x_1 : y_2 - y_1 : z_2 - z_1$$

Angle θ between two lines, whose direction angles are $\alpha_1, \beta_1, \gamma_1$ and $\alpha_2, \beta_2, \gamma_2$:

$$\cos \theta = \cos \alpha_1 \cos \alpha_2 + \cos \beta_1 \cos \beta_2 + \cos \gamma_1 \cos \gamma_2$$

Equation of a plane is of the first degree in x, y, and z:

$$Ax + By + Cz + D = 0$$

where A, B, C are proportional to the direction cosines of a normal or perpendicular to the plane.

Angle between two planes is the angle between their normals. Equations of a straight line are two equations of the first degree.

$$A_1 x + B_1 y + C_1 z + D_1 = 0 \qquad A_2 x + B_2 y + C_2 z + D_2 = 0$$

Equations of a straight line through the point $P_1(x_1, y_1, z_1)$ with direction cosines proportional to a, b, and c:

$$\frac{x - x_1}{a} = \frac{y - y_1}{b} = \frac{z - z_1}{c}$$

Differential Calculus

Derivative Relations

If $x = f(y)$, then $\dfrac{dy}{dx} = 1 \div \dfrac{dx}{dy}$

If $x = f(t)$ and $y = F(t)$, then $\dfrac{dy}{dx} = \dfrac{dy}{dt} \div \dfrac{dx}{dt}$

If $y = f(u)$ and $u = F(x)$, then $\dfrac{dy}{dx} = \dfrac{dy}{du} \cdot \dfrac{du}{dx}$

Differential Calculus, continued

Derivatives

Functions of x represented by u and v, constants by a, n, and e:

$$\frac{d}{dx}(x) = 1 \qquad \frac{d}{dx}(a) = 0$$

$$\frac{d}{dx}(u \pm v \pm \cdots) = \frac{du}{dx} \pm \frac{dv}{dx} \pm \cdots$$

$$\frac{d}{dx}(au) = a\frac{du}{dx} \qquad \frac{d}{dx}a^u = a^u \ln a \frac{du}{dx}$$

$$\frac{d}{dx}(uv) = u\frac{dv}{dx} + v\frac{du}{dx} \qquad \frac{d}{dx}e^u = e^u \frac{du}{dx}$$

$$\frac{d}{dx}\left(\frac{u}{v}\right) = \left(v\frac{du}{dx} - u\frac{dv}{dx}\right)v^2 \qquad \frac{d}{dx}u^v = vu^{v-1}\frac{du}{dx} + u^v \ln u \frac{dv}{dx}$$

$$\frac{d}{dx}(u^n) = nu^{n-1}\frac{du}{dx} \qquad \frac{d}{dx}\sin u = \cos u \frac{du}{dx}$$

$$\frac{d}{dx}\log_a u = \frac{\log_a e}{u}\frac{du}{dx}$$

$$\frac{d}{dx}\ln u = \frac{1}{u}\frac{du}{dx}$$

$$\frac{d}{dx}\cos u = -\sin u \frac{du}{dx} \qquad \frac{d}{dx}\tan u = \sec^2 u \frac{du}{dx}$$

$$\frac{d}{dx}\cot u = -\csc^2 u \frac{du}{dx} \qquad \frac{d}{dx}\sin^{-1} u = \frac{1}{\sqrt{1-u^2}}\frac{du}{dx}$$

(where $\sin^{-1} u$ lies between $-\pi/2$ and $+\pi/2$)

$$\frac{d}{dx}\cos^{-1} u = -\frac{1}{\sqrt{1-u^2}}\frac{du}{dx}$$

(where $\cos^{-1} u$ lies between 0 and π)

$$\frac{d}{dx}\tan^{-1} u = \frac{1}{1+u^2}\frac{du}{dx} \qquad \frac{d}{dx}\cot^{-1} u = -\frac{1}{1+u^2}\frac{du}{dx}$$

$$\frac{d}{dx}\sec^{-1} u = \frac{1}{u\sqrt{u^2-1}}\frac{du}{dx}$$

(where $\sec^{-1} u$ lies between 0 and π)

$$\frac{d}{dx}\csc^{-1} u = -\frac{1}{u\sqrt{u^2-1}}\frac{du}{dx}$$

(where $\csc^{-1} u$ lies between $-\pi/2$ and $+\pi/2$)

Differential Calculus, continued

nth Derivative of Certain Functions

$$\frac{d^n}{dx^n} e^{ax} = a^n e^{ax}$$

$$\frac{d^n}{dx^n} a^x = (\ln a)^n a^x$$

$$\frac{d^n}{dx^n} \ln x = \frac{(-1)^{n-1} \, \lfloor n-1}{x^n} \qquad \lfloor n-1 = 1 \cdot 2 \cdot 3 \cdots (n-1)$$

$$\frac{d^n}{dx^n} \sin ax = a^n \sin\left(ax + \frac{n\pi}{2}\right)$$

$$\frac{d^n}{dx^n} \cos ax = a^n \cos\left(ax + \frac{n\pi}{2}\right)$$

Integral Calculus

Theorems on Integrals and Some Examples

$$\int d f(x) = f(x) + C$$

$$d \int f(x)\,dx = f(x)\,dx$$

$$\int \left[f_1(x) \pm f_2(x) \pm \cdots \right] dx = \int f_1(x)\,dx \pm \int f_2(x)\,dx \pm \cdots$$

$$\int a f(x)\,dx = a \int f(x)\,dx \quad a \text{ any constant}$$

$$\int u^n\,du = \frac{u^{n+1}}{n+1} + C \quad (n \neq -1) \quad u \text{ any function of } x$$

$$\int \frac{du}{u} = \ln u + C \quad u \text{ any function of } x$$

$$\int u\,dv = uv - \int v\,du \quad u \text{ and } v \text{ any functions of } x$$

Integral Calculus, continued

Selected Integrals

$$\int du = u + C$$

$$\int a \, du = a \int du = au + C$$

$$\int [f_1(u) + f_2(u) + \cdots + f_n(u)] \, du = \int f_1(u) \, du + \int f_2(u) \, du + \int f_n(u) \, du$$

$$\int u^n \, du = \frac{u^{n+1}}{n+1} + C \quad (n \neq -1)$$

$$\int \frac{du}{u} = \ln |u| + C$$

$$\int a^u \, du = \frac{a^u}{\ln a} + C \quad (a > 0, a \neq 1)$$

$$\int e^u \, du = e^u + C$$

$$\int \sin u \, du = -\cos u + C$$

$$\int \cos u \, du = \sin u + C$$

$$\int \sec^2 u \, du = \tan u + C$$

$$\int \csc^2 u \, du = -\cot u + C$$

$$\int \sec u \tan u \, du = \sec u + C$$

$$\int \csc u \cot u \, du = -\csc u + C$$

$$\int \tan u \, du = -\ln |\cos u| + C = \ln |\sec u| + C$$

Integral Calculus, continued

$$\int \cot u \ du = \ell_n |\sin u| + C$$

$$\int \sec u \ du = \ell_n |\sec u + \tan u| + C$$

$$\int \csc u \ du = \ell_n |\csc u - \cot u| + C$$

$$\int \frac{du}{\sqrt{a^2 - u^2}} = \arcsin \frac{u}{a} + C \quad (a > 0)$$

$$\int \frac{du}{a^2 + u^2} = \frac{1}{a} \arctan \frac{u}{a} + C$$

$$\int u \ dv = uv - \int v \ du$$

$$\int \frac{du}{\sqrt{u^2 \pm a^2}} = \ell_n \left| u + \sqrt{u^2 \pm a^2} \right| + C$$

$$\int \frac{du}{u\sqrt{u^2 - a^2}} = \frac{1}{a} \text{arcsec} \frac{u}{a} + C \quad (a > 0)$$

$$\int \sqrt{a^2 - u^2} \ du = \frac{u}{2}\sqrt{a^2 - u^2} + \frac{a^2}{2}\arcsin\frac{u}{a} + C \quad (a > 0)$$

$$\int \sqrt{u^2 \pm a^2} \ du = \frac{u}{2}\sqrt{u^2 \pm a^2} \pm \frac{a^2}{2}\ell_n \left| u + \sqrt{u^2 \pm a^2} \right| + C$$

Some Definite Integrals

$$\int_0^a \sqrt{a^2 - x^2} \ dx = \frac{\pi a^2}{4} \qquad \int_0^a \sqrt{2ax - x^2} \ dx = \frac{\pi a^2}{4}$$

$$\int_0^\infty \frac{dx}{ax^2 + b} = \frac{\pi}{2\sqrt{ab}} \qquad a \text{ and } b \text{ positive}$$

$$\int_0^\pi \sin^2 ax \ dx = \int_0^\pi \cos^2 ax \ dx = \frac{\pi}{2}$$

$$\int_0^\infty e^{-ax^2} \ dx = \frac{1}{2}\sqrt{\frac{\pi}{a}}$$

$$\int_0^\infty x^n e^{-ax} \ dx = \frac{n!}{a^{n-1}} \qquad n \text{ a positive integer}$$

Differential Equations

Equations of First Order and First Degree: $Mdx + Ndy = 0$

Variables Separable

$$X_1 Y_1 \, dx + X_2 Y_2 \, dy = 0$$

$$\int \frac{X_1}{X_2} \, dx + \int \frac{Y_2}{Y_1} \, dy = C$$

Homogeneous Equation

$$dy - f\left(\frac{y}{x}\right) dx = 0$$

$$x = Ce \int^{[dv/f(v)-v]} \quad \text{and} \quad v = \frac{y}{x}$$

$M \div N$ can be written so that x and y occur only as $y \div x$, if every term in M and N have the same degree in x and y.

Linear Equation

$$dy + (X_1 y - X_2) \, dx = 0$$

$$y = e^{-\int x_1 \, dx}\left(\int X_2 e^{\int X_1 dx} dx + C\right)$$

A similar solution exists for $dx + (Y_1 x - Y_2) \, dy = 0$.

Exact Equation

$$Mdx + Ndy = 0 \quad \text{where } \partial M/\partial y = \partial N/\partial x$$

$$\int Mdx + \int \left|N - \frac{\partial}{\partial y}\int Mdx\right| dy = C$$

for y constant when integrating with respect to x.

Nonexact Equation

$$Mdx + Ndy = 0 \quad \text{where } \partial M/\partial y \neq \partial N/\partial x$$

Make equation exact by multiplying by an integrating factor $\mu(x, y)$—a form readily recognized in many cases.

Complex Quantities

Properties of Complex Quantities

If z, z_1, z_2 represent complex quantities, then

Sum or difference: $z_1 \pm z_2 = (x_1 \pm x_2) + j(y_1 \pm y_2)$

Product: $z_1 \cdot z_2 = r_1 r_2 [\cos(\theta_1 + \theta_2) + j \sin(\theta_1 + \theta_2)]$

$$= r_1 r_2 e^{j(\theta_1 + \theta_2)} = (x_1 x_2 - y_1 y_2) + j(x_1 y_2 + x_2 y_1)$$

Quotient: $z_1 / z_2 = r_1 / r_2 [\cos(\theta_1 - \theta_2) + j \sin(\theta_1 - \theta_2)]$

$$= \frac{r_1}{r_2} e^{j(\theta_1 - \theta_2)} = \frac{x_1 x_2 + y_1 y_2}{x_2^2 + y_2^2} + j \frac{x_2 y_1 - x_1 y_2}{x_2^2 + y_2^2}$$

Power: $z^n = r^n [\cos n\theta + j \sin n\theta] = r^n e^{jn\theta}$

Root: $\sqrt[n]{z} = \sqrt[n]{r} \left| \cos \frac{\theta + 2k\pi}{n} + j \sin \frac{\theta + 2k\pi}{n} \right| = \sqrt[n]{r} e^{j(\theta + 2k\pi/n)}$

where k takes in succession the values $0, 1, 2, 3, \ldots, n - 1$.
If $z_1 = z_2$, then $x_1 = x_2$ and $y_1 = y_2$.

Periodicity: $z = r(\cos \theta + j \sin \theta)$

$$= r[\cos(\theta + 2k\pi) + j \sin(\theta + 2k\pi)]$$

or

$$z = r e^{j\theta} = r e^{j(\theta + 2k\pi)} \quad \text{and} \quad e^{j2k\pi} = 1$$

where k is any integer.

Exponential-trigonometric relations:

$$e^{jz} = \cos z + j \sin z \qquad e^{-jz} = \cos z - j \sin z$$

$$\cos z = \frac{1}{2}(e^{jz} + e^{-jz}) \qquad \sin z = \frac{1}{2j}(e^{jz} - e^{-jz})$$

Some Standard Series

Binomial Series

$$(a + x)^n = a^n + na^{n-1}x + \frac{n(n-1)}{2!}a^{n-2}x^2$$

$$+ \frac{n(n-1)(n-2)}{3!}a^{n-3}x^3 + \cdots \quad (x^2 < a^2)$$

The number of terms becomes infinite when n is negative or fractional.

$$(a - bx)^{-1} = \frac{1}{a}\left(1 + \frac{bx}{a} + \frac{b^2 x^2}{a^2} + \frac{b^3 x^3}{a^3} + \cdots\right) \quad (b^2 x^2 < a^2)$$

Some Standard Series, continued

Exponential Series

$$a^x = 1 + x \ln a + \frac{(x \ln a)^2}{2!} + \frac{(x \ln a)^3}{3!} + \cdots$$

$$e^x = 1 + x + \frac{x^2}{2!} + \frac{x^3}{3!} + \cdots$$

Logarithmic Series

$$\ln x = (x - 1) - \tfrac{1}{2}(x - 1)^2 + \tfrac{1}{3}(x - 1)^3 - \cdots \quad (0 < x < 2)$$

$$\ln x = \frac{x - 1}{x} + \frac{1}{2}\left(\frac{x - 1}{x}\right)^2 + \frac{1}{3}\left(\frac{x - 1}{x}\right)^3 + \cdots \quad \left(x > \frac{1}{2}\right)$$

$$\ln x = 2\left[\frac{x - 1}{x + 1} \cdot \frac{1}{3}\left(\frac{x - 1}{x + 1}\right)^3 + \frac{1}{5}\left(\frac{x - 1}{x + 1}\right)^5 + \cdots\right] \quad (x \text{ positive})$$

$$\ln(1 + x) = x - \frac{x^2}{2} + \frac{x^3}{3} - \frac{x^4}{4} + \cdots$$

Trigonometric Series

$$\sin x = x - \frac{x^3}{3!} + \frac{x^5}{5!} - \frac{x^7}{7!} + \cdots$$

$$\cos x = 1 - \frac{x^2}{2!} + \frac{x^4}{4!} - \frac{x^6}{6!} + \cdots$$

$$\tan x = x + \frac{x^3}{3} + \frac{2x^5}{15} + \frac{17x^7}{315} + \frac{62x^9}{2835} + \cdots \quad \left(x^2 < \frac{\pi^2}{4}\right)$$

$$\sin^{-1} x = x + \frac{1}{2}\frac{x^3}{3} + \frac{1 \cdot 3}{2 \cdot 4}\frac{x^5}{5} + \frac{1 \cdot 3 \cdot 5}{2 \cdot 4 \cdot 6}\frac{x^7}{7} + \cdots \quad (x^2 < 1)$$

$$\tan^{-1} x = x - \frac{1}{3}x^3 + \frac{1}{5}x^5 - \frac{1}{7}x^7 + \cdots \quad (x^2 \leq 1)$$

Some Standard Series, continued

Other Series

$$1 + 2 + 3 + 4 + \cdots + (n - 1) + n = n(n + 1)/2$$

$$p + (p + 1) + (p + 2) + \cdots + (q - 1) + q = (q + p)(q - p + 1)/2$$

$$2 + 4 + 6 + 8 + \cdots + (2n - 2) + 2n = n(n + 1)$$

$$1 + 3 + 5 + 7 + \cdots + (2n - 3) + (2n - 1) = n^2$$

$$1^2 + 2^2 + 3^3 + 4^2 + \cdots + (n - 1)^2 + n^2 = n(n + 1)(2n + 1)/6$$

$$1^3 + 2^3 + 3^3 + 4^3 + \cdots + (n - 1)^3 + n^3 = n^2(n + 1)^2/4$$

$$\left. \begin{array}{c} \dfrac{1 + 2 + 3 + 4 + 5 + \cdots + n}{n^2} \rightarrow \dfrac{1}{2} \\[2ex] \dfrac{1 + 2^2 + 3^2 + 4^2 + \cdots + n^2}{n^3} \rightarrow \dfrac{1}{3} \\[2ex] \dfrac{1 + 2^3 + 3^3 + 4^3 + \cdots + n^3}{n^4} \rightarrow \dfrac{1}{4} \end{array} \right\} \text{ as } n \rightarrow \infty$$

Matrix Operations

Two matrices $\mathbf{A}$ and $\mathbf{B}$ can be added if the number of rows in $\mathbf{A}$ equals the number of rows in $\mathbf{B}$.

$$\mathbf{A} \pm \mathbf{B} = \mathbf{C}$$

$$a_{ij} \pm b_{ij} = c_{ij} \qquad i = 1, 2, \ldots, m \qquad j = 1, 2, \ldots, n$$

Multiplying a matrix or vector by a scalar implies multiplication of each element by the scalar. If $\mathbf{B} = \gamma \mathbf{A}$, then $b_{ij} = \gamma a_{ij}$ for all elements.

Two matrices, $\mathbf{A}$ and $\mathbf{B}$, can be multiplied if the number of columns in $\mathbf{A}$ equals rows in $\mathbf{B}$. For $\mathbf{A}$ of order $m \times n$ (m rows and n columns) and $\mathbf{B}$ of order $n \times p$, the product of two matrices $\mathbf{C} = \mathbf{AB}$ will be a matrix of order $m \times p$ elements

$$c_{ij} = \sum_{k=1}^{n} a_{ik} n_{kj}$$

Thus, c_{ij} is the scalar product of the i'th row vector of $\mathbf{A}$ and the j'th column vector of $\mathbf{B}$.

Matrix Operations, continued

In general, matrix multiplication is not commutative: $\mathbf{AB} \neq \mathbf{BA}$. Matrix multiplication is associative: $\mathbf{A(BC)} = \mathbf{(AB)C}$. The distributive law for multiplication and addition holds as in the case of scalars.

$$(\mathbf{A} + \mathbf{B})\mathbf{C} = \mathbf{AC} + \mathbf{BC}$$

$$\mathbf{C}(\mathbf{A} + \mathbf{B}) = \mathbf{CA} + \mathbf{CB}$$

For some applications, the term-by-term product of two matrices $\mathbf{A}$ and $\mathbf{B}$ of identical order is defined as $\mathbf{C} = \mathbf{A} \cdot \mathbf{B}$, where $c_{ij} = a_{ij}b_{ij}$.

$$(\mathbf{ABC})' = \mathbf{C}' \mathbf{B}' \mathbf{A}'$$

$$(\mathbf{ABC})^H = \mathbf{C}^H \mathbf{B}^H \mathbf{A}^H$$

If both $\mathbf{A}$ and $\mathbf{B}$ are symmetric, then $(\mathbf{AB})' = \mathbf{BA}$. The product of two symmetric matrices will usually not be symmetric.

Determinants

A determinant $|\mathbf{A}|$ or $\det(\mathbf{A})$ is a scalar function of a square matrix.

$$|\mathbf{A}||\mathbf{B}| = |\mathbf{AB}|$$

$$\begin{vmatrix} a_{11} & a_{12} \\ a_{21} & a_{22} \end{vmatrix} = a_{11}a_{22} - a_{12}a_{21}$$

$$|\mathbf{A}| = |\mathbf{A}'|$$

$$\begin{vmatrix} a_{11} & a_{12} & a_{13} \\ a_{21} & a_{22} & a_{23} \\ a_{31} & a_{32} & a_{33} \end{vmatrix} = \begin{array}{l} a_{11}a_{22}a_{33} + a_{12}a_{23}a_{31} + a_{13}a_{21}a_{32} \\ - a_{13}a_{22}a_{31} - a_{11}a_{23}a_{32} - a_{12}a_{21}a_{33} \end{array}$$

$$\begin{vmatrix} a_{11} & a_{12} & \cdots & a_{1n} \\ a_{21} & a_{22} & \cdots & a_{2n} \\ \cdot & \cdot & \cdots & \cdot \\ a_{n1} & a_{n2} & \cdots & a_{nm} \end{vmatrix} = \sum (-1)^{\delta} a_{1i}, a_{2i}, \ldots, a_{ni_n}$$

where the sum is over all permutations: $i_1 \neq i_2 \neq \cdots \neq i_n$, and δ denotes the number of exchanges necessary to bring the sequence $(i_1, i_2, \ldots, i_n)$ back into the natural order $(1, 2, \ldots, n)$.

Curve Fitting

Polynomial Function

$$y = b_0 + b_1 x + b_2 x^2 + \cdots + b_m x^m$$

For a polynomial function fit by the method of least squares, obtain the values of $b_0, b_1, \ldots, b_m$ by solving the system of $m + 1$ normal equations.

$$n b_0 + b_1 \Sigma x_i + b_2 \Sigma x_i^2 + \cdots + b_m \Sigma x_i^m = \Sigma y_i$$

$$b \Sigma x_i + b_1 \Sigma x_i^2 + b_2 \Sigma x_i^3 + \cdots + b_m \Sigma x_i^{m+1} = \Sigma x_i y_i$$

$$b_0 \Sigma x_i^m + b_1 \Sigma x_i^{m+1} + b_2 \Sigma x_i^{m+2} + \cdots + b_m \Sigma x_i^{2m} = \Sigma x_i^m y_i$$

Straight Line

$$y = b_0 + b_1 x$$

For a straight line fit by the method of least squares, obtain the values b_0 and b_1 by solving the normal equations.

$$n b_0 + b_1 \Sigma x_i = \Sigma y_i$$

$$b_0 \Sigma x_i + b_1 \Sigma x_i^2 = \Sigma x_i y_i$$

Solutions for these normal equations:

$$b_1 = \frac{n \Sigma x_i y_i - (\Sigma x_i)(\Sigma y_i)}{n \Sigma x_i^2 - (\Sigma x_i)^2} \qquad b_0 = \frac{\Sigma y_i}{n} - b_1 \frac{\Sigma x_i}{n} = \bar{y} - b_1 \bar{x}$$

Exponential Curve

$$y = ab^x \quad \text{or} \quad \log y = \log a + (\log b)x$$

For an exponential curve fit by the method of least squares, obtain the values $\log a$ and $\log b$ by fitting a straight line to the set of ordered pairs $\{(x_i, \log y_i)\}$.

Power Function

$$y = ax^b \quad \text{or} \quad \log y = \log a + b \log x$$

For a power function fit by the method of least squares, obtain the values $\log a$ and b by fitting a straight line to the set or ordered pairs $\{(\log x_i, \log y_i)\}$.

Small-Term Approximations

This section lists some first approximations derived by neglecting all powers but the first of the small positive or negative quantity, $x = s$. The expression in brackets gives the next term beyond that used, and, by means of it, the accuracy of the approximation can be estimated.

$$\frac{1}{1+s} = 1 - s \qquad [+s^2]$$

$$(1+s)^n = 1 + ns \qquad \left[+\frac{n(n-1)}{2}s^2\right]$$

$$e^s = 1 + s \qquad \left[+\frac{s^2}{2}\right]$$

$$\ln(1+s) = s \qquad \left[-\frac{s^2}{2}\right]$$

$$\sin s = s \qquad \left[-\frac{s^3}{6}\right]$$

$$\cos s = 1 \qquad \left[-\frac{s^2}{2}\right]$$

$$(1+s_1)(1+s_2) = (1+s_1+s_2) \qquad [+s_1 s_2]$$

The following expressions may be approximated by $1 + s$, where s is a small positive or negative quantity and n any number:

$$e^s \qquad 2 - e^{-s} \qquad \cos\sqrt{-2s}$$

$$\left(1+\frac{s}{n}\right)^n \qquad 1 + \ln\sqrt{\frac{1+s}{1-s}}$$

$$\sqrt[n]{1+ns} \qquad 1 + n\sin\frac{s}{n}$$

$$\sqrt[n]{\frac{1+ns/2}{1-ns/2}} \qquad 1 + n\ln\left(1+\frac{s}{n}\right)$$

Vector Equations

A vector is a line segment that has both magnitude (length) and direction. Examples are velocity, acceleration, and force. A unit vector has length one. A scalar quantity has only a magnitude, e.g., mass, temperature, and energy.

Properties of Vectors

A vector may be expressed in terms of a set of unit vectors along specified directions and magnitudes of components in those directions. For example, using a set of mutually perpendicular unit vectors (i, j, k). $V = ai + bj$ in two dimensions, or $V = ai + bj + ck$ in three dimensions.

The vector sum V of any number of vectors V_1, V_2, V_3, where $V_1 = a_1 i + b_1 j + c_1 k$, etc., is given by

$$V = V_1 + V_2 + V_3 + \cdots = (a_1 + a_2 + a_3 + \cdots) i$$

$$+ (b_1 + b_2 + b_3 + \cdots) j + (c_1 + c_2 + c_3 + \cdots) k$$

Product of a Vector V by a Scalar Quantity s

$$sV = (sa)i + (sb)j + (sc)k$$

$$(s_1 + s_2)V = s_1 V + s_2 V \qquad (V_1 + V_2)s = V_1 s + V_2 s$$

where sV has the same direction as V, and its magnitude is s times the magnitude of V.

Scalar Product of Two Vectors, V₁ · V₂

$$V_1 \cdot V_2 = |V_1||V_2| \cos \phi$$

where ϕ is the angle between V_1 and V_2.

$$V_1 \cdot V_1 = |V_1|^2 \quad \text{and} \quad i \cdot i = j \cdot j = k \cdot k = 1$$

because $\cos \phi = \cos(0) = 1$. On the other hand,

$$i \cdot j = j \cdot k = k \cdot i = 0$$

because $\cos \phi = \cos(90°) = 0$.

Scalar products behave much like products in normal (scalar) arithmetic.

$$V_1 \cdot V_2 = V_2 \cdot V_1 \qquad (V_1 + V_2) \cdot V_3 = V_1 \cdot V_3 + V_2 \cdot V_3$$

In a plane,

$$V_1 \cdot V_2 = a_1 a_2 + b_1 b_2$$

Vector Equations, continued

In space,

$$V_1 \cdot V_2 = a_1 a_2 + b_1 b_2 + c_1 c_2$$

An example of the scalar product of two vectors $V_1 \cdot V_2$ is the work done by a constant force of magnitude $|V_1|$ acting through a distance $|V_2|$, where ϕ is the angle between the line of force and the direction of motion.

Vector Product of Two Vectors, $V_1 \times V_2$

$$|V_1 \times V_2| = |V_1||V_2| \sin \phi$$

where ϕ is the angle between V_1 and V_2.

Among unit vectors, the significance of the order of vector multiplication can be seen.

$$i \times j = k \qquad j \times k = i \qquad k \times i = j$$

but reversing the order changes the sign, so that

$$j \times i = -k, \text{etc.}$$

and

$$V_1 \times V_2 = -V_2 \times V_1$$

Also

$$i \times i = j \times j = k \times k = 0$$

and in general

$$V_1 \times V_1 = 0$$

because $\sin \phi = \sin(0) = 0$.

An example of a vector product is the moment M_0 about a point 0 of a force F_P applied at point P, where $r_{P/0}$ is the position vector of the point of force application with respect to 0.

$$M_0 = r_{P/0} \times F_P$$

Changing the order of multiplication would change the sign of the moment, which is equivalent to switching between clockwise and counterclockwise moments.

Section 2

SECTION PROPERTIES

Plane Areas

Figure	General properties	Moment of inertia		Radius of gyration, P
		Area	Weight	

Square

General properties:

Area: S^2

Centroid: $\bar{x} = \bar{y} = \dfrac{S}{2}$

Moment of inertia — Area:
$$I_x = I_y = \frac{S^4}{12}$$
$$I_{x_1} = I_{y_1} = \frac{S^4}{3}$$
$$I_p = I_x + I_y = \frac{S^4}{6}$$
$$I_{P_1} = I_{x_1} + I_{y_1} = \frac{2S^4}{3}$$

Moment of inertia — Weight:
$$I_x = I_y = \frac{WS^2}{12}$$
$$I_{x_1} = I_{y_1} = \frac{WS^2}{3}$$
$$I_p = I_x + I_y = \frac{WS^2}{6}$$
$$I_{P_1} = I_{x_1} + I_{y_1} = \frac{2WS^2}{3}$$

Radius of gyration, P:
$$P_x = P_y = 0.289S$$
$$P_{x_1} = P_{y_1} = 0.577S$$
$$P_p = 0.408S$$
$$P_{P_1} = 0.816S$$

Rectangle

General properties:

Area: BH

Centroid: $\bar{x} = \dfrac{B}{2}$, $\bar{y} = \dfrac{H}{2}$

Moment of inertia — Area:
$$I_x = \frac{BH^3}{12} \qquad I_y = \frac{HB^3}{12}$$
$$I_p = \frac{BH}{12}(H^2 + B^2)$$
$$I_{x_1} = \frac{BH^3}{3}$$
$$I_{y_1} = \frac{HB^3}{3}$$
$$I_{P_1} = \frac{BH}{3}(H^2 + B^2)$$

Moment of inertia — Weight:
$$I_x = \frac{WH^2}{12} \qquad I_y = \frac{WB^2}{12}$$
$$I_p = \frac{W}{12}(H^2 + B^2)$$
$$I_{x_1} = \frac{WH^2}{3} \qquad I_{y_1} = \frac{WB^2}{3}$$
$$I_{P_1} = \frac{W}{3}(H^2 + B^2)$$

Radius of gyration, P:
$$P_x = 0.289H$$
$$P_y = 0.289B$$
$$P_p = 0.289\sqrt{H^2 + B^2}$$
$$P_{x_1} = 0.577H$$
$$P_{y_1} = 0.577B$$
$$P_{P_1} = 0.577\sqrt{H^2 + B^2}$$

Plane Areas, continued

Figure	General properties	Moment of inertia		Radius of gyration, P
		Area	Weight	

Hollow square

General properties:

Area:
$$S^2 - s^2$$

Centroid:
$$\bar{x} = \bar{y} = \frac{S}{2}$$

Area:
$$I_x = I_y = \frac{S^4 - s^4}{12}$$
$$I_p = \frac{S^4 - s^4}{6}$$
$$I_{x_1} = I_{y_1} = \frac{4S^4 - 3S^2 s^2 - s^4}{12}$$
$$I_{p_1} = \frac{4S^4 - 3S^2 s^2 - s^4}{6}$$

Weight:
$$I_x = I_y = \frac{W(S^2 + s^2)}{12}$$
$$I_{x_1} = I_{y_1} = \frac{W(4S^2 + s^2)}{12}$$
$$I_p = \frac{W(S^2 + s^2)}{6}$$
$$I_{p_1} = \frac{W(4S^2 + s^2)}{6}$$

Radius of gyration:
$$P_x = P_y = \sqrt{\frac{S^2 + s^2}{12}}$$
$$P_{x_1} = P_{y_1} = \sqrt{\frac{(4S^2 + s^2)}{12}}$$
$$P_p = 0.408\sqrt{S^2 + s^2}$$
$$P_{p_1} = \sqrt{\frac{(4S^2 + s^2)}{6}}$$

Hollow rectangle

General properties:

Area:
$$BH - bh$$

Centroid:
$$\bar{x} = \frac{B}{2}$$
$$\bar{y} = \frac{H}{2}$$

Area:
$$I_x = \frac{BH^3 - bh^3}{12}$$
$$I_y = \frac{HB^3 - hb^3}{12}$$
$$I_{x_1} = \frac{BH^3}{3} - \frac{bh(3H^2 + h^2)}{12}$$
$$I_{y_1} = I_y + \frac{(BH - bh)B^2}{4}$$
$$I_p = I_x + I_y$$
$$I_{p_1} = I_{x_1} + I_{y_1}$$

Weight:
$$I_x = \frac{W}{12}\left[\frac{BH^3 - bh^3}{BH - bh}\right]$$
$$I_y = \frac{W}{12}\left[\frac{HB^3 - hb^3}{BH - bh}\right]$$
$$I_{x_1} = I_x + \frac{WH^2}{4}$$
$$I_{y_1} = I_y + \frac{WB^2}{4}$$
$$I_p = I_x + I_y$$
$$I_{p_1} = I_{x_1} + I_{y_1}$$

Radius of gyration:
$$P_x = \sqrt{\frac{BH^3 - bh^3}{12(BH - bh)}}$$
$$P_y = \sqrt{\frac{HB^3 - hb^3}{12(BH - bh)}}$$
$$P_{x_1} = \sqrt{\frac{I_{x_1}}{BH - bh}}$$
$$P_{y_1} = \sqrt{\frac{I_{y_1}}{BH - bh}}$$
$$P_{p_1} = \sqrt{\frac{I_{p_1}}{BH - bh}}$$
$$P_p = \sqrt{\frac{I_p}{BH - bh}}$$

Plane Areas, continued

Figure	General properties	Moment of inertia		Radius of gyration, P
		Area	Weight	
Oblique triangle	Area: $\dfrac{1}{2}BH$ Centroid: $\bar{x} = \dfrac{B+C}{3}$ $\bar{y} = \dfrac{H}{3}$	$I_x = \dfrac{BH^3}{36}$ $I_{x_1} = \dfrac{BH^3}{12}$ $I_{x_2} = \dfrac{BH^3}{4}$ $I_y = \dfrac{BH}{36}(B^2 + C^2 - BC)$ $I_p = \dfrac{BH}{36}(H^2 + B^2 + C^2 - BC)$	$I_x = \dfrac{WH^2}{18}$ $I_{x_1} = \dfrac{WH^2}{6}$ $I_{x_2} = \dfrac{WH^2}{2}$ $I_y = \dfrac{W}{18}(B^2 + C^2 - BC)$ $I_p = \dfrac{W}{18}(H^2 + B^2 + C^2 - BC)$	$P_x = 0.236H$ $P_{x_1} = 0.408H$ $P_{x_2} = 0.707H$ $P_y = 0.236\sqrt{B^2 + C^2 - BC}$ $P_p = 0.236\sqrt{H^2 + B^2 + C^2 - BC}$
Isoceles trapezoid	Area: $\dfrac{H(A+B)}{2}$ Centroid: $\bar{Y}_a = \dfrac{H(B+2A)}{3(B+A)}$ $\bar{Y}_b = \dfrac{H(A+2B)}{3(A+B)}$	$I_x = \dfrac{H^3(A^2 + 4AB + B^2)}{36(A+B)}$ $I_{x_1} = \dfrac{H^3(A+B)(2A+C)}{12(A+C)}$ $I_y = \dfrac{H(A+B)(A^2+B^2)}{48}$ $I_{y_1} = \dfrac{H(A+B)(A^2+7B^2)}{48}$ $I_p = I_x + I_y$ $I_{p_1} = I_{x_1} + I_{y_1}$	$I_x = \dfrac{WH^2}{18}\left[1 + \dfrac{2AB}{(A+B)^2}\right]$ $I_{x_1} = \dfrac{WH^2(3A+B)}{6(A+B)}$ $I_y = \dfrac{W}{24}(A^2+B^2)$ $I_{y_1} = \dfrac{W}{24}(A^2+7B^2)$ $I_p = I_x + I_y$	$P_x = H\dfrac{\sqrt{2(A^2+4AB+B^2)}}{6(A+B)}$ $P_y = 0.204\sqrt{A^2+B^2}$ $P_p = \sqrt{\dfrac{2I_p}{H(A+B)}}$

Plane Areas, continued

Figure	General properties	Moment of inertia — Area	Moment of inertia — Weight	Radius of gyration, P
Regular polygon $\theta = \dfrac{180°}{n}$ $B = 2\sqrt{R^2 - R_1^2}$ n = number of sides	Area: $$\frac{nB^2 \cot\theta}{4}$$ $$\frac{nR^2 \sin 2\theta}{2}$$ $$nR_1^2 \tan\theta$$ Centroid: $$\bar{x} = \bar{y} = 0$$	$$I_y = I_N = \frac{A(6R^2 - B^2)}{24}$$ $$= \frac{A(12R_1^2 + B^2)}{48}$$	$$R = \frac{B}{2\sin\theta} \quad R_1 = \frac{B}{2\tan\theta}$$ $$I_y = I_N = \frac{m(6R^2 - B^2)}{24}$$ $$= \frac{m(12R_1^2 + B^2)}{48}$$	$$P_y = P_N = \sqrt{\frac{6R^2 - B^2}{24}}$$ $$= \sqrt{\frac{12R_1^2 + B^2}{48}}$$
Regular hexagon	Area: $0.866H^2$ Centroid: $$\bar{x} = \frac{B}{2} = A$$ $$\bar{y} = \frac{H}{2}$$	$I_x = I_y = 0.0601H^4$ $I_{x_1} = 0.2766H^4$ $I_{y_1} = 0.3488H^4$ $I_p = 0.1203H^4$	$I_x = I_y = 0.0694WH^2$ $= 0.0521WB^2$ $I_{x_1} = 0.3194WH^2$ $I_p = 0.1389WH^2 - 0.1042WB^2$ $I_{y_1} = 0.4028WH^2 = 0.3021WB^2$	$P_x = P_y = 0.2635H$ $= 0.2282B$ $P_{x_1} = 0.5652H$ $P_{y_1} = 0.6346H$ $P_p = 0.3727H$

Plane Areas, continued

Figure	General properties	Moment of inertia — Area	Moment of inertia — Weight	Radius of gyration, P

Right angled trapezoid

General properties

Area:
$$\frac{H}{2}(2A+B)$$

Centroid:
$$\bar{x} = \frac{3A^2 + 3AB + B^2}{3(2A+B)}$$

$$\bar{y} = \frac{H(3A+B)}{3(2A+B)}$$

Moment of inertia — Area

$$I_x = \frac{H^3(6A^2 + 6AB + B^2)}{36(2A+B)}$$

$$I_{x_1} = \frac{H^3(4A+B)}{12}$$

$$I_y = I_{y_1} - \frac{H(3A^2 + 3AB + B^2)^2}{18(2A+B)}$$

$$I_{y_1} = \frac{H}{12}(2A+B) \times (2A^2 + 2AB + B^2)$$

$$I_p = I_x + I_y$$

$$I_{p_1} = I_{x_1} + I_{y_1}$$

Moment of inertia — Weight

$$I_x = \frac{WH^2(6A^2 + 6AB + B^2)}{18(2A+B)^2}$$

$$I_{x_1} = \frac{WH^2(4A+B)}{6(2A+B)}$$

$$I_y = I_{y_1} - W\bar{X}^2$$

$$I_{y_1} = \frac{W}{6}(2A^2 + 2AB + B^2)$$

$$I_p = I_x + I_y$$

$$I_{p_1} = I_{x_1} + I_{y_1}$$

Radius of gyration, P

$$P_x = \frac{0.236H}{(2A+B)}\sqrt{6A^2 + 6AB + B^2}$$

$$P_{x_1} = 0.408H\sqrt{\frac{4A+B}{2A+B}}$$

$$P_y = \sqrt{\frac{2I_y}{H(2A+B)}}$$

$$P_{y_1} = \sqrt{\frac{2A^2 + 2AB + B^2}{6}}$$

$$P_p = \sqrt{\frac{2I_p}{H(2A+B)}}$$

Oblique trapezoid

General properties

Area:
$$\frac{1}{2}H(A+B)$$

Centroid:
x is on a line connecting midpoints of sides A and B.

$$\bar{y}_a = \frac{H(B+2A)}{3(B+A)}$$

$$\bar{y}_b = \frac{H(A+2B)}{3(A+B)}$$

Moment of inertia — Area

$$I_x = \frac{H^3(A^2 + 4AB + B^2)}{36(A+B)}$$

$$I_{x_1} = \frac{H^3(B+3A)}{12}$$

Moment of inertia — Weight

$$I_x = \frac{WH^2}{18}\left(1 + \frac{2AB}{(A+B)^2}\right)$$

$$I_{x_1} = \frac{WH^2}{6}\left(\frac{3A+B}{A+B}\right)$$

Radius of gyration, P

$$P_x = H\sqrt{\frac{2(A^2 + 4AB + B^2)}{6(A+B)}}$$

$$P_{x_1} = H\sqrt{\frac{3A+B}{6(A+B)}}$$

Plane Areas, continued

Figure	General properties	Moment of inertia		Radius of gyration, P
		Area	Weight	
Parallelogram	Area: $$BH$$ Centroid: $$\bar{x} = \frac{A+B}{2}$$ $$\bar{y} = \frac{H}{2}$$	$I_x = \frac{BH^3}{12}$ $I_{x_1} = \frac{BH^3}{3}$ $I_y = \frac{BH(A^2+B^2)}{12}$ $I_{y_1} = \frac{BH}{6}(2A^2+2B^2+3AB)$ $I_p = \frac{BH}{12}(A^2+B^2+H^2)$ $I_{p_1} = \frac{BH}{6}(2A^2+2B^2+3AB+2H^2)$	$I_x = \frac{WH^2}{12}$ $I_{x_1} = \frac{WH^2}{3}$ $I_y = \frac{W}{12}(A^2+B^2)$ $I_{y_1} = \frac{W}{6}(2A^2+2B^2+3AB)$ $I_p = \frac{W}{12}(A^2+B^2+H^2)$ $I_{p_1} = \frac{W(2A^2+2B^2+3AB+2H^2)}{6}$	$P_x = 0.289H$ $P_{x_1} = 0.577H$ $P_y = 0.289\sqrt{A^2+B^2}$ $P_{y_1} = 0.408\sqrt{2A^2+2B^2+3AB}$ $P_p = 0.289\sqrt{A^2+B^2+H^2}$ $P_{p_1} = 0.408\sqrt{2A^2+2B^2+3AB+2H^2}$
Regular octagon	Area: $$2.8284R^2$$ Centroid: $$\bar{x} = \bar{y} = R$$	$I_x = I_y = 0.6381R^4$ $I_{x_1} = I_{y_1} = 3.4665R^4$ $I_p = 1.2761R^4$	$I_x = I_y = 0.2256WR^2$ $I_{x_1} = I_{y_1} = 1.2256WR^2$ $I_p = 0.4512WR^2$	$P_x = P_y = 0.4750R$ $P_{x_1} = P_{y_1} = 1.1071R$ $P_p = 0.672R$

Plane Areas, continued

Figure	General properties	Moment of inertia		Radius of gyration, P
		Area	Weight	
Circle	Area: $$0.7854D^2$$ Centroid: $$\bar{x} = \bar{y} = R$$	$I_x = I_y = 0.0491D^4$ $I_{x_1} = I_{y_1} = 0.2454D^4$ $I_p = 0.0982D^4$	$I_x = I_y = \dfrac{WD^2}{16} = \dfrac{WR^2}{4}$ $I_{x_1} = I_{y_1} = 1.25WR^2$ $I_p = \dfrac{WD^2}{8} = \dfrac{WR^2}{2}$	$P_x = P_y = \dfrac{D}{4}$ $P_{x_1} = P_{y_1} = 0.5590D = 1.118R$ $P_p = 0.3536D$
Hollow circle	Area: $$\pi(R^2 - r^2)$$ Centroid: $$\bar{x} = \bar{y} = R$$	$I_x = I_y = \dfrac{\pi(R^4 - r^4)}{4}$ $I_p = \dfrac{\pi(R^4 - r^4)}{2}$ $I_{x_1} = I_{y_1} = \dfrac{\pi(5R^4 - 4R^2r^2 - r^4)}{4}$	$I_x = I_y = \dfrac{W(R^2 + r^2)}{4}$ $I_p = \dfrac{W(R^2 + r^2)}{2}$ $I_{x_1} = I_{y_1} = \dfrac{W(5R^2 + r^2)}{4}$	$P_x = P_y = \dfrac{1}{2}\sqrt{R^2 + r^2}$ $P_p = \sqrt{\dfrac{R^2 + r^2}{2}}$ $P_{x_1} = I_{y_1} = \dfrac{1}{2}\sqrt{5R^2 + r^2}$

Plane Areas, continued

Figure	General properties	Moment of inertia — Area	Moment of inertia — Weight	Radius of gyration, P
Semicircle 	Area: $0.3927D^2 = 1.571R^2$ Centroid: $\bar{x} = R$ $\bar{y} = 0.2122D = 0.4244R$ $= \dfrac{4R}{3\pi}$	$I_x = 0.1098R^4$ $I_{x_1} = 0.3927R^4$ $I_y = 0.3927R^4$ $I_{y_1} = 1.9635R^4$ $I_p = 0.5025R^4$ $I_{p_1} = 2.3562R^4$	$I_x = 0.06987WR^2$ $I_{x_1} = 0.25WR^2$ $I_y = 0.25WR^2$ $I_{y_1} = 1.25WR^2$ $I_p = 0.3199WR^2$ $I_{p_1} = 1.50WR^2$	$P_x = 0.264R = 0.132D$ $P_{x_1} = 0.5R = 0.25D$ $P_y = 0.5R = 0.25D$ $P_{y_1} = 1.118R = 0.559D$ $P_p = 0.566R = 0.2828D$ $P_{p_1} = 1.225R = 0.6124D$
Hollow semicircle 	Area: $\dfrac{\pi(R^2 - r^2)}{2}$ Centroid: $\bar{x} = R$ $\bar{y} = 0.4244\left(R + \dfrac{r^2}{R+r}\right)$	$I_x = \dfrac{\pi}{8}(R^4 - r^4) - \dfrac{\pi(R^2 - r^2)}{2}\bar{y}^2$ $I_{x_1} = \dfrac{\pi}{8}(R^4 - r^4)$ $I_y = \dfrac{\pi}{8}(R^4 - r^4)$ $I_{y_1} = \dfrac{\pi(R^2 - r^2)(5R^2 + r^2)}{8}$ $I_p = I_x + I_y$ $I_{p_1} = I_{x_1} + I_{y_1}$	$I_x = \dfrac{W(R^2 + r^2)}{4} - W\bar{y}^2$ $I_y = \dfrac{W(R^2 + r^2)}{4}$ $I_p = I_x + I_y$ $I_{p_1} = \dfrac{W(3R^2 + r^2)}{2}$ $I_{x_1} = \dfrac{W(R^2 + r^2)}{4}$ $I_{y_1} = \dfrac{W(5R^2 + r^2)}{4}$	$P_x = \sqrt{\dfrac{2I_x}{\pi(R^2 - r^2)}}$ $P_{x_1} = \sqrt{\dfrac{R^2 + r^2}{4}}$ $P_y = \sqrt{\dfrac{R^2 + r^2}{4}}$ $P_{y_1} = \sqrt{\dfrac{2I_{y_1}}{\pi(R^2 - r^2)}}$ $P_p = \sqrt{\dfrac{2I_p}{\pi(R^2 - r^2)}}$

Plane Areas, continued

Figure	General properties	Moment of inertia		
		Area	Weight	Radius of gyration, P
Ellipse	Area: πAB Centroid: $\bar{x} = A$ $\bar{y} = B$	$I_x = \dfrac{\pi AB^3}{4} = 0.7854\,AB^3$ $I_{x_1} = 1.25 - AB^3$ $I_y = \dfrac{\pi A^3 B}{4} = 0.7854\,A^3 B$ $I_{y_1} = 1.25 - A^3 B$ $I_p = \dfrac{\pi AB(A^2 + B^2)}{4}$	$I_x = \dfrac{WB^2}{4}$ $I_{x_1} = 1.25\,WB^2$ $I_y = \dfrac{WA^2}{4}$ $I_{y_1} = 1.25\,WA^2$ $I_p = \dfrac{W(A^2 + B^2)}{4}$	$P_x = \dfrac{B}{2}$ $P_{x_1} = 1.118\,B$ $P_y = \dfrac{A}{2}$ $P_{y_1} = 1.118\,A$ $P_p = \sqrt{\dfrac{A^2 + B^2}{2}}$
Hollow ellipse	Area: $\pi(AB - CD)$ Centroid: $\bar{x} = A$ $\bar{y} = B$	$I_x = \dfrac{\pi}{4}(AB^3 - CD^3)$ $I_{x_1} = \dfrac{\pi}{4}(AB^3 - CD^3)$ $\quad + \pi(AB - CD)(B^2)$ $I_y = \dfrac{\pi}{4}(A^3 B - C^3 D)$ $I_{y_1} = \dfrac{\pi}{4}(A^3 B - C^3 D)$ $\quad + \pi(AB - CD)(A^2)$ $I_p = I_x + I_y$	$I_x = \dfrac{W}{4}\left[\dfrac{AB^3 - CD^3}{AB - CD}\right]$ $I_{x_1} = \dfrac{W}{4}\left[\dfrac{AB^3 - CD^3}{AB - CD}\right] + WB^2$ $I_y = \dfrac{W}{4}\left[\dfrac{A^3 B - C^3 D}{AB - CD}\right]$ $I_{y_1} = \dfrac{W}{4}\left[\dfrac{A^3 B - C^3 D}{AB - CD}\right] + WA^2$ $I_p = I_x + I_y$	$P_x = \sqrt{\dfrac{AB^3 - CD^3}{4(AB - CD)}}$ $P_{x_1} = \sqrt{\dfrac{I_{x_1}}{\pi(AB - CD)}}$ $P_y = \sqrt{\dfrac{A^3 B - C^3 D}{4(AB - CD)}}$ $P_{y_1} = \sqrt{\dfrac{I_{y_1}}{\pi(AB - CD)}}$ $P_p = \sqrt{\dfrac{I_p}{\pi(AB - CD)}}$

Plane Areas, continued

Figure	General properties	Moment of inertia — Area	Moment of inertia — Weight	Radius of gyration, P
Semiellipse	Area: $$\frac{\pi AB}{2}$$ Centroid: $$\bar{x} = A$$ $$\bar{y} = 0.424B$$	$I_x = 0.1098AB^3$ $I_{x_1} = 0.3927AB^3$ $I_y = 0.3927A^3B$ $I_{y_1} = 1.9635A^3B$	$I_x = 0.070WB^2$ $I_{x_1} = 0.25WB^2$ $I_y = 0.25WA^2$ $I_{y_1} = 1.25WA^2$ $I_p = \dfrac{W(A^2 + 0.280B^2)}{4}$ $I_{p_1} = \dfrac{W(5A^2 + B^2)}{4}$	$P_x = 0.2643B$ $P_{x_1} = \dfrac{B}{2}$ $P_y = \dfrac{A}{2}$ $P_{y_1} = 1.118A$ $P_p = \sqrt{\dfrac{A^2}{4} + \dfrac{B^2}{4} - \dfrac{16B^2}{9\pi^2}}$ $P_{p_1} = \sqrt{\dfrac{2I_{p_1}}{\pi AB}}$
Hollow semiellipse	Area: $$\frac{\pi(AB - CD)}{2}$$ Centroid: $$\bar{x} = A$$ $$\bar{y} = \frac{4}{3\pi}\left[\frac{AB^2 - CD^2}{AB - CD}\right]$$	$I_x = \dfrac{\pi}{8}(AB^3 - CD^3)$ $\quad - \dfrac{\pi(AB - CD)}{2}\left[\dfrac{4(AB^2 - CD^2)}{3\pi(AB - CD)}\right]^2$ $I_{x_1} = \dfrac{\pi}{8}(AB^3 - CD^3)$ $I_y = \dfrac{\pi}{8}(A^3B - C^3D)$ $I_{y_1} = \dfrac{\pi}{8}(A^3B - C^3D)$ $\quad + \dfrac{\pi A^2(AB - CD)}{2}$	$I_x = \dfrac{W}{4}\left(\dfrac{AB^3 - CD^3}{AB - CD}\right)$ $\quad - W\left[\left(\dfrac{4}{3\pi}\right)\left(\dfrac{AB^2 - CD^2}{AB - CD}\right)\right]^2$ $I_{x_1} = \dfrac{W}{4}\left(\dfrac{AB^3 - CD^3}{AB - CD}\right)$ $I_y = \dfrac{W}{4}\left(\dfrac{A^3B - C^3D}{AB - CD}\right)$ $I_{y_1} = \dfrac{W}{4}\left(\dfrac{A^3B - C^3D}{AB - CD}\right) + WA^2$ $I_p = I_x + I_y$	$P_x = \sqrt{\dfrac{2I_x}{\pi(AB - CD)}}$ $P_{x_1} = \sqrt{\dfrac{AB^3 - CD^3}{4(AB - CD)}}$ $P_y = \sqrt{\dfrac{2I_y}{\pi(AB - CD)}}$ $P_{y_1} = \sqrt{\dfrac{2I_{y_1}}{\pi(AB - CD)}}$ $P_p = \sqrt{\dfrac{2I_p}{\pi(AB - CD)}}$

Plane Areas, continued

Figure	General properties	Moment of inertia		Radius of gyration, P
		Area	Weight	

Circular sector

a in radians

Area:
$$R^2 a$$

Centroid:
$$\bar{x} = \frac{2}{3}\left[\frac{R\sin a}{a}\right]$$
$$\bar{y} = R\sin a$$

Moment of inertia — Area:
$$I_x = \frac{R^4}{4}(a - \sin a \cos a)$$
$$I_y = \frac{R^4}{4}\left(a - \frac{16\sin^2 a}{9a} + \frac{\sin 2a}{2}\right)$$
$$I_{y_1} = \frac{R^4}{4}(a + \sin a \cos a)$$
$$I_p = \frac{R^4}{4}\left(2a - \frac{16\sin^2 a}{9a}\right)$$

Moment of inertia — Weight:
$$I_x = \frac{WR^2}{4a}(a - \sin a \cos a)$$
$$I_{x_1} = I_x + WR^2\sin^2 a$$
$$I_y = \frac{WR^2}{4a}\left(a - \frac{16\sin^2 a}{9a} + \frac{\sin 2a}{2}\right)$$
$$I_{y_1} = \frac{WR^2}{4a}(a + \sin a \cos a)$$
$$I_p = I_x + I_y$$

Radius of gyration, P:
$$P_x = \frac{R}{2}\sqrt{1 - \frac{\sin a \cos a}{a}}$$
$$P_{x_1} = \sqrt{\frac{I_{x_1}}{R^2 a}}$$
$$P_y = \frac{R}{2}\sqrt{1 + \frac{\sin a \cos a}{a} - \frac{16\sin^2 a}{9a^2}}$$
$$P_{y_1} = \frac{R}{2}\sqrt{1 + \frac{\sin a \cos a}{a}}$$
$$P_p = \frac{R}{2}\sqrt{2 - \frac{16\sin^2 a}{9a^2}}$$

Hollow circular sector

a in radians

Area:
$$(R^2 - r^2)a$$

Centroid:
$$\bar{x} = \frac{2\sin a(R^3 - r^3)}{3a(R^2 - r^2)}$$
$$\bar{y} = R\sin a$$

Moment of inertia — Area:
$$I_x = \frac{a}{4}(R^4 - r^4)\times\left(1 - \frac{\sin a \cos a}{a}\right)$$
$$I_y = I_{y_1} - \frac{1}{a(R^2 - r^2)}\left[\frac{2\sin a(R^3 - r^3)}{3}\right]^2$$
$$I_{y_1} = \frac{a}{4}(R^4 - r^4)\times\left(1 + \frac{\sin a \cos a}{a}\right)$$

Moment of inertia — Weight:
$$I_x = \frac{W}{4}(R^2 + r^2)\times\left(1 - \frac{\sin a \cos a}{a}\right)$$
$$I_{x_1} = I_x + WR^2\sin^2 a$$
$$I_y = I_{y_1} - W\left[\frac{2\sin a(R^3 - r^3)}{3a(R^2 - r^2)}\right]^2$$
$$I_{y_1} = \frac{W}{4}(R^2 + r^2)\left(1 + \frac{\sin a \cos a}{a}\right)$$
$$I_p = I_x + I_y$$

Radius of gyration, P:
$$P_x = \sqrt{\frac{R^2 + r^2}{4}\left[1 - \frac{\sin a \cos a}{a}\right]}$$
$$P_{x_1} = \sqrt{\frac{I_{x_1}}{(R^2 - r^2)a}}$$
$$P_y = \sqrt{\frac{I_y}{(R^2 - r^2)a}}$$
$$P_{y_1} = \sqrt{\frac{R^2 + r^2}{4}\left[1 + \frac{\sin a \cos a}{a}\right]}$$
$$P_p = \sqrt{\frac{I_p}{(R^2 - r^2)a}}$$

Plane Areas, continued

Figure	General properties	Moment of inertia — Area	Moment of inertia — Weight	Radius of gyration, P
Circular segment *a in radians*	Area: $\dfrac{R^2}{2}(2a - \sin 2a)$ Centroid: $\bar{x} = \dfrac{4R\sin^3 a}{3(2a - \sin 2a)}$ $\bar{y} = R\sin a$	$I_x = \dfrac{AR^2}{4}\left[1 - \dfrac{2\sin^3 a\cos a}{3(a - \sin a\cos a)}\right]$ $I_y = I_{y_1} - \dfrac{4R^6\sin^6 a}{9A}$ $I_{y_1} = \dfrac{AR^2}{4}\left[1 + \dfrac{2\sin^3 a\cos a}{a - \sin a\cos a}\right]$	$I_x = \dfrac{WR^2}{4}\left[1 - \dfrac{2\sin^3 a\cos a}{3(a - \sin a\cos a)}\right]$ $I_{x_1} = I_x + WR^2\sin^2 a$ $I_y = I_{y_1} - W\left[\dfrac{2R\sin^3 a}{3(a - \sin a\cos a)}\right]^2$ $I_{y_1} = \dfrac{W^2}{4}\left[1 + \dfrac{2\sin^3 a\cos a}{a - \sin a\cos a}\right]$ $I_p = I_x + I_y$	$P_x = \sqrt{\dfrac{R^2}{4}\left[1 - \dfrac{2\sin^3 a\cos a}{3(a - \sin a\cos a)}\right]}$ $P_y = \sqrt{\dfrac{2I_y}{R^2(2a \cdot \sin 2a)}}$ $P_p = \sqrt{\dfrac{2I_p}{R^2(2a \cdot \sin 2a)}}$
Parabolic segment	Area: $\dfrac{4}{3}AB$ Centroid: $\bar{x} = 0.6A$ $\bar{y} = B$	$I_x = 0.2667AB^3$ $I_{x_1} = 1.6AB^3$ $I_y = 0.0914A^3B$ $I_{y_1} = 0.5714A^3B$ $I_p = I_x + I_y$	$I_x = 0.2WB^2$ $I_{x_1} = 1.2WB^2$ $I_y = 0.0686WA^2$ $I_{y_1} = 0.4286WA^2$ $I_p = I_x + I_y$	$P_x = 0.4472B$ $P_{x_1} = 1.095B$ $P_y = 0.2619A$ $P_{y_1} = 0.6547A$ $P_p = \sqrt{\dfrac{3I_p}{4AB}}$

Plane Areas, continued

Figure	General properties	Moment of inertia		Radius of gyration, P
		Area	Weight	
Parabolic half-segment	Area: $\dfrac{2AB}{3}$ Centroid: $\bar{x} = 0.6A$ $\bar{y} = 0.375B$	$I_x = 0.0396AB^3$ $I_{x_1} = 0.1333AB^3$ $I_y = 0.0457A^3B$ $I_{y_1} = 0.2857A^3B$ $I_p = I_x + I_y$ $I_{p_1} = I_{x_1} + I_{y_1}$	$I_x = 0.0594W \cdot B^2$ $I_{x_1} = 0.2W \cdot B^2$ $I_y = 0.0686W A^2$ $I_{y_1} = 0.4286W A^2$ $I_p = I_x + I_y$ $I_{p_1} = I_{x_1} + I_{y_1}$	$P_x = 0.2437B$ $P_{x_1} = 0.4472B$ $P_y = 0.2619A$ $P_{y_1} = 0.6547A$ $P_p = \sqrt{\dfrac{3I_p}{2AB}}$ $P_{p_1} = \sqrt{\dfrac{3I_{p_1}}{2AB}}$
Nose rib	Area: $\dfrac{2}{3}A(B+C)$ Centroid: $\bar{x} = 0.6A$ $\bar{y} = 0.375(B - C)$	$I_x = \dfrac{A(B+C)}{480}$ $\times (19B^2 + 26BC + 19C^2)$ $I_y = 0.0457A^3(B+C)$ $I_{y_1} = 0.2857A^3(B+C)$ $I_p = I_x + I_y$ $I_{p_1} = I_{x_1} + I_{y_1}$	$I_x = \dfrac{W}{320}(19B^2 + 26BC + 19C^2)$ $I_{x_1} = \dfrac{W}{5}(B^2 - BC + C^2)$ $I_y = 0.0686W A^2$ $I_{y_1} = 0.4286W A^2$ $I_p = I_x + I_y$ $I_{p_1} = I_{x_1} + I_{y_1}$	$P_x = \sqrt{\dfrac{3I_x}{2A(B+C)}}$ $P_y = \sqrt{\dfrac{3I_y}{2A(B+C)}}$

based on parabolic segment

Plane Areas, continued

Figure	General properties	Moment of inertia		Radius of gyration, P
		Area	Weight	

Equilateral triangle

General properties:

Area: $\dfrac{BH}{2}$

Centroid:
$\bar{x} = \dfrac{B}{2}$
$\bar{y} = \dfrac{H}{3}$

Area moment of inertia:
$$I_x = I_y = \frac{B^3 H}{48} \qquad I_{y_1} = \frac{7B^3 H}{48}$$
$$I_{x_1} = \frac{B^3 H}{16} \qquad I_p = \frac{B^3 H}{24}$$
$$I_{x_2} = \frac{BH^3}{4} \qquad I_{p_1} = \frac{5B^3 H}{24}$$

Weight:
$$I_x = I_y = \frac{WB^2}{24} \qquad I_{y_1} = \frac{7WB^2}{24}$$
$$I_{x_1} = \frac{WB^2}{8} \qquad I_p = \frac{WB^2}{12}$$
$$I_{x_2} = \frac{WH^2}{2} \qquad I_{p_1} = \frac{5WB^2}{12}$$

Radius of gyration:
$$P_x = 0.204B$$
$$P_{x_1} = 0.354B$$
$$P_{x_2} = 0.707H$$
$$P_y = 0.204B$$
$$P_{y_1} = 0.540B$$
$$P_p = 0.289B$$
$$P_{p_1} = 0.645B$$

Rhombus

General properties:

Area: BH

Centroid:
$\bar{y} = \dfrac{H}{2}$
$\bar{x} = \dfrac{A+B}{2}$

Area moment of inertia:
$$I_x = \frac{BH^3}{12}$$
$$I_{x_1} = \frac{BH^3}{3}$$
$$I_y = \frac{BH(A^2 + B^2)}{12}$$
$$I_{y_1} = \frac{BH(2A^2 + 2B^2 + 3AB)}{6}$$
$$I_p = \frac{1}{6}B^3 H$$
$$I_{p_1} = \frac{B^2 H(3A + 4B)}{6}$$

Weight:
$$I_x = \frac{WH^2}{12}$$
$$I_{x_1} = \frac{WH^2}{3}$$
$$I_y = \frac{W(A^2 + B^2)}{12}$$
$$I_{y_1} = \frac{W(2A^2 + 2B^2 + 3AB)}{6}$$
$$I_p = \frac{WB^2}{6}$$
$$I_{p_1} = \frac{WB(3A + 4B)}{6}$$

Radius of gyration:
$$P_x = 0.289H$$
$$P_{x_1} = 0.577H$$
$$P_y = 0.289\sqrt{(A^2 + B^2)}$$
$$P_{y_1} = 0.408\sqrt{2A^2 + 2B^2 + 3AB}$$
$$P_p = 0.408B$$
$$P_{p_1} = 0.408\sqrt{B(3A + 4B)}$$

Plane Areas, continued

Figure	General properties	Moment of inertia		Radius of gyration, P
		Area	Weight	
Obtuse angled triangle	Area: $$\frac{BH}{2}$$ Centroid: $$\bar{x} = \frac{B+2C}{3}$$ $$\bar{y} = \frac{H}{3}$$	$$I_x = \frac{BH^3}{36}$$ $$I_{x_1} = \frac{BH^3}{12}$$ $$I_{x_2} = \frac{BH^3}{4}$$ $$I_y = \frac{BH}{36}(B^2+BC+C^2)$$ $$I_{y_1} = \frac{BH}{12}(B^2+3BC+3C^2)$$ $$I_p = \frac{BH}{36}(H^2+B^2+BC+C^2)$$	$$I_x = \frac{WH^2}{18}$$ $$I_{x_1} = \frac{WH^2}{6}$$ $$I_{x_2} = \frac{WH^2}{2}$$ $$I_y = \frac{W}{18}(B^2+BC+C^2)$$ $$I_{y_1} = \frac{W(B^2+3BC+3C^2)}{6}$$ $$I_p = \frac{W(H^2+B^2+BC+C^2)}{18}$$	$$P_x = 0.236H$$ $$P_{x_1} = 0.408H$$ $$P_{x_2} = 0.707H$$ $$P_y = 0.236\sqrt{B^2+BC+C^2}$$ $$P_{y_1} = 0.408\sqrt{B^2+3BC+3C^2}$$ $$P_p = 0.236\sqrt{H^2+B^2+BC+C^2}$$

Solids

Figure	General properties	Moment of inertia	Radius of gyration, P
Cube	Volume: $$A^3$$ Centroid: $$\bar{x} = \bar{y} = \bar{z} = \frac{A}{2}$$	$$I_x = I_y = I_z = \frac{WA^2}{6}$$ $$I_{x_1} = I_{y_1} = I_{z_1} = \frac{2WA^2}{3}$$ $$I_{x_2} = \frac{5WA^2}{12}$$	$$P_x = P_y = P_z = 0.408A$$ $$P_{x_1} = P_{y_1} = P_{z_1} = 0.816A$$ $$P_{x_2} = 0.646A$$
Rectangular prism	Volume: $$ABH$$ Centroid: $$\bar{x} = \frac{A}{2}$$ $$\bar{y} = \frac{B}{2}$$ $$\bar{z} = \frac{H}{2}$$	$$I_x = \frac{W}{12}(B^2 + H^2)$$ $$I_y = \frac{W}{12}(A^2 + H^2)$$ $$I_z = \frac{W}{12}(A^2 + B^2)$$	$$P_x = 0.289\sqrt{B^2 + H^2}$$ $$P_y = 0.289\sqrt{A^2 + H^2}$$ $$P_z = 0.289\sqrt{A^2 + B^2}$$
Sphere	Volume: $$\frac{4}{3}\pi R^3$$ Area: $$4\pi R^2$$	$$I_x = I_y = I_z = \frac{2}{5}WR^2$$	$$P_x = P_y = P_z = 0.632R$$

Solids, continued

Figure	General properties	Moment of inertia	Radius of gyration, P
Hollow sphere	Volume: $\frac{4}{3}\pi(R^3 - r^3)$	$I_x = I_y = I_z = \frac{2}{5}W\left(\frac{R^5 - r^5}{R^3 - r^3}\right)$	$P_x = P_y = P_z = 0.632\sqrt{\left(\frac{R^5 - r^5}{R^3 - r^3}\right)}$
Hemisphere 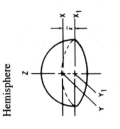	Volume: $\frac{2}{3}\pi R^3$ Centroid: $\bar{z} = \frac{3}{8}R$	$I_x = I_y = 0.26WR^2$ $I_{x_1} = I_{y_1} = I_z = 0.4WR^2$	$P_x = P_y = 0.51R$ $P_{x_1} = P_{y_1} = P_z = 0.632R$

Solids, continued

Figure	General properties	Moment of inertia	Radius of gyration, P
Right circular cylinder	Volume: $\pi R^2 H$ Centroid: $\bar{z} = \dfrac{H}{2}$	$I_x = I_y = \dfrac{W}{12}(3R^2 + H^2)$ $I_{x_1} = I_{y_1} = \dfrac{W}{12}(3R^2 + 4H^2)$ $I_z = \dfrac{WR^2}{2}$	$P_x = P_y = 0.289\sqrt{3R^2 + H^2}$ $P_{x_1} = P_{y_1} = 0.289\sqrt{3R^2 + 4H^2}$ $P_z = 0.707R$
Hollow right circular cylinder	Volume: $\pi H(R^2 - r^2)$ Centroid: $\bar{z} = \dfrac{H}{2}$	$I_x = I_y = \dfrac{W}{12}[3(R^2 + r^2) + H^2]$ $I_{x_1} = I_{y_1} = W\left[\dfrac{R^2 + r^2}{4} + \dfrac{H^2}{3}\right]$ $I_z = \dfrac{W}{2}(R^2 + r^2)$	$P_x = P_y = 0.289\sqrt{3(R^2 + r^2) + H^2}$ $P_z = 0.707\sqrt{R^2 + r^2}$ $P_{x_1} = P_{y_1} = \sqrt{\dfrac{R^2 + r^2}{4} + \dfrac{H^2}{3}}$

Solids, continued

Figure	General properties	Moment of inertia	Radius of gyration, P
Elliptical cylinder 	Volume: πABH Centroid: $\bar{z} = \dfrac{H}{2}$	$I_x = \dfrac{W}{12}(3B^2 + H^2)$ $I_y = \dfrac{W}{12}(3A^2 + H^2)$ $I_z = \dfrac{W}{4}(A^2 + B^2)$	$P_x = 0.289\sqrt{3B^2 + H^2}$ $P_y = 0.289\sqrt{3A^2 + H^2}$ $P_z = \dfrac{\sqrt{A^2 + B^2}}{2}$
Ellipsoid $A > B > C$	If $C = B$, then resulting solid is a prolate spheroid (i.e., ellipse rotated about major axis). Volume: $\dfrac{4}{3}\pi ABC$ If $A = B$, then resulting solid is an oblate spheroid (i.e., ellipse rotated about minor axis).	$I_x = \dfrac{W}{5}(A^2 + C^2)$ $I_{x_1} = \dfrac{W}{5}(6A^2 + C^2)$ $I_y = \dfrac{W}{5}(A^2 + B^2)$ $I_{y_1} = \dfrac{W}{5}(6A^2 + B^2)$ $I_z = \dfrac{W}{5}(B^2 + C^2)$	$P_x = 0.447\sqrt{A^2 + C^2}$ $P_{x_1} = 0.447\sqrt{6A^2 + C^2}$ $P_y = 0.447\sqrt{A^2 + B^2}$ $P_{y_1} = 0.447\sqrt{6A^2 + B^2}$ $P_z = 0.447\sqrt{B^2 + C^2}$

Solids, continued

Figure	General properties	Moment of inertia	Radius of gyration, P
Paraboloid of revolution	Volume: $\dfrac{\pi R^2 H}{2}$ Centroid: $\bar{z} = \dfrac{H}{3}$	$I_x = I_y = \dfrac{W}{18}(3R^2 + H^2)$ $I_{x_1} = \dfrac{W}{6}(R^2 + H^2)$ $I_{x_2} = \dfrac{W}{6}(R^2 + 3H^2)$ $I_z = \dfrac{WR^2}{3}$	$P_x = P_y = 0.236\sqrt{3R^2 + H^2}$ $P_{x_1} = 0.408\sqrt{R^2 + H^2}$ $P_{x_2} = 0.408\sqrt{R^2 + 3H^2}$ $P_z = 0.577R$
Solid elliptical hemispheroid	Volume: $2/3\pi R^2 H$ Centroid: $\bar{z} = 3/8H$	$I_x = I_y = W\left[\dfrac{R^2}{5} + \dfrac{19H^2}{320}\right]$ $I_z = 2/5WR^2$	$P_x = P_y = 0.4475\sqrt{R^2 + 0.297H^2}$ $P_z = 0.632R$

Solids, continued

Figure	General properties	Moment of inertia	Radius of gyration, P
Right rectangular pyramid 	Volume: $\dfrac{ABH}{3}$ Centroid: $\bar{z} = \dfrac{H}{4}$	$I_x = \dfrac{W}{20}\left(B^2 + \dfrac{3H^2}{4}\right)$ $I_{x_1} = \dfrac{W}{20}(B^2 + 2H^2)$ $I_y = \dfrac{W}{20}\left(A^2 + \dfrac{3}{4}H^2\right)$ $I_{y_1} = \dfrac{W}{20}(A^2 + 2H^2)$ $I_z = \dfrac{W}{20}(A^2 + B^2)$	$P_x = 0.224\sqrt{B^2 + \dfrac{3H^2}{4}}$ $P_y = 0.224\sqrt{A^2 + \dfrac{3H^2}{4}}$ $P_z = 0.224\sqrt{A^2 + B^2}$
Right angled wedge 	Volume: $\dfrac{ABH}{2}$ Centroid: $\bar{x} = \dfrac{A}{3}$ $\bar{y} = \dfrac{B}{2}$ $\bar{z} = \dfrac{H}{3}$	$I_x = \dfrac{W}{36}(2H^2 + 3B^2)$ $I_y = \dfrac{W}{18}(A^2 + H^2)$ $I_z = \dfrac{W}{36}(2A^2 + 3B^2)$	$P_x = 0.167\sqrt{2H^2 + 3B^2}$ $P_y = 0.236\sqrt{A^2 + H^2}$ $P_z = 0.167\sqrt{2A^2 + 3B^2}$

Solids, continued

Figure	General properties	Moment of inertia	Radius of gyration, P

Right circular cone

General properties:

Volume:
$$\frac{\pi R^2 H}{3}$$

Centroid:
$$\bar{z} = \frac{H}{4}$$

Moment of inertia:

$$I_x = I_y = \frac{3W}{20}\left(R^2 + \frac{H^2}{4}\right)$$

$$I_{x_1} = I_{y_1} = \frac{W}{20}(3R^2 + 2H^2)$$

$$I_z = \frac{3W}{10}R^2$$

$$I_{x_2} = \frac{3W}{20}(R^2 + 4H^2)$$

Radius of gyration, P:

$$P_x = P_y = 0.387\sqrt{R^2 + \frac{H^2}{4}}$$

$$P_{x_1} = P_{y_1} = 0.224\sqrt{3R^2 + 2H^2}$$

$$P_{x_2} = 0.387\sqrt{R^2 + 4H^2}$$

$$P_z = 0.548R$$

Frustum of a cone

General properties:

Volume:
$$\frac{\pi H}{3}(R^2 + Rr + r^2)$$

Centroid:
$$\bar{z} = \frac{H}{4}\left[\frac{R^2 + 2Rr + 3r^2}{R^2 + Rr + r^2}\right]$$

Moment of inertia:

$$I_{x_1} = I_{y_1} = \frac{WH^2}{10}\left[\frac{R^2 + 3Rr + 6r^2}{R^2 + Rr + r^2}\right] + \frac{3W}{10}\left[\frac{R^5 - r^5}{R^3 - r^3}\right]$$

$$I_z = \frac{3W}{10}\left[\frac{R^5 - r^5}{R^3 - r^3}\right]$$

$$I_x = I_y = I_{x_1} - W(z)^2$$

Radius of gyration, P:

$$P_z = 0.548\sqrt{\frac{R^5 - r^5}{R^3 - r^3}}$$

Solids, continued

Figure	General properties	Moment of inertia	Radius of gyration, P
Spherical sector	Volume: $$\frac{2}{3}\pi R^2 H$$ Centroid: $$\bar{z} = \frac{3}{8}(2R - H)$$	$$I_z = \frac{WH}{5}(3R - H)$$	$$P_z = 0.447\sqrt{(3R - H)(H)}$$
Spherical segment	Volume: $$\frac{\pi H}{6}(3r^2 + H^2)$$ $$= \frac{\pi H^2}{3}(3R - H)$$ Centroid: $$\bar{z} = \frac{3}{4}\frac{(2R - H)^2}{(3R - H)}$$ Area: $$2\pi RH$$	$$I_z = \frac{2HW}{3R - H}\left(R^2 - \frac{3}{4}RH + \frac{3}{20}H^2\right)$$	$$P_z = \sqrt{\frac{2H}{3R - H}\left(R^2 - \frac{3}{4}RH + \frac{3}{20}H^2\right)}$$

Solids, continued

Figure	General properties	Moment of inertia	Radius of gyration, P
Elliptic paraboloid	Volume: $$\frac{\pi A B H}{2}$$ Centroid: $$\bar{z} = \frac{H}{3}$$	$$I_x = \frac{W}{18}(3B^2 + H^2)$$ $$I_{x_1} = \frac{W}{6}(B^2 + H^2)$$ $$I_y = \frac{W}{18}(3A^2 + H^2)$$ $$I_{y_1} = \frac{W}{6}(A^2 + H^2)$$ $$I_z = \frac{W}{6}(A^2 + B^2)$$	$$P_x = 0.236\sqrt{3B^2 + H^2}$$ $$P_y = 0.236\sqrt{3A^2 + H^2}$$ $$P_z = 0.408\sqrt{A^2 + B^2}$$ $$P_{x_1} = 0.408\sqrt{B^2 + H^2}$$ $$P_{y_1} = 0.408\sqrt{A^2 + H^2}$$
Torus	Volume: $$2\pi^2 r^2 R$$ Centroid: $$\bar{x} = \bar{z} = R + r$$ $$\bar{y} = r$$	$$I_x = I_z = \frac{W}{8}(4R^2 + 5r^2)$$ $$I_y = \frac{W}{4}(4R^2 + 3r^2)$$	$$P_x = P_z = 0.354\sqrt{4R^2 + 5r^2}$$ $$P_y = \frac{\sqrt{4R^2 + 3r^2}}{2}$$
Outer half of solid torus	Length: $$2\pi R$$ Volume: $$\frac{4}{3}\pi r^3 + \pi^2 r^2 R$$	$$I_x = \frac{W}{4.189r + 9.870R}[1.6755r^3 + 7.4022Rr^2 + 12.5664R^2r + 9.869R^3]$$ $$I_y = I_z = \frac{W}{12}[27\pi Rr^2 + 36R^2r + 44r^3]$$	

Solids, continued

Figure	General properties	Moment of inertia

Prismoid

$\bar{x} = A_0$ to C.G.

A_0 = area of base

A_1 = area of top

A_m = area at $h/2$

Note: A_0, A_1, and A_m must be parallel.

Volume:

$$\frac{h}{6}(A_0 + A_1 + 4A_m)$$

Centroid:

$$\bar{X} = \frac{h(A_1 + 2A_m)}{(A_0 + A_1 + 4A_m)}$$

$$I = \frac{\rho h}{20}\left(3I_0 + 16I_m + 3I_1 - \frac{A_1 I_0}{A_0} - \frac{A_0 I_i}{A_i}\right)$$

where:

$A_0 = b_0 c_0$ ； $A_1 = b_1 c_1$

$$A_m = \left[\frac{b_0 + b_1}{2}\right]\left[\frac{c_0 + c_1}{2}\right]$$

I_0 = area moment of inertia of A_0

I_m = area moment of inertia of A_m

I_1 = area moment of inertia of A_1

Solid ogive

R = ogive radius

d = radius of truncated nose

　　($d = 0$ for a complete ogive)

$D = R - r$

$\sigma = r - d$

$$V = \frac{\pi h}{9}\left[(3r - a)^2 + 2a^2 - \frac{1.2a^2(R - r)}{R}\right]$$

Complete ogive (1/2 football):

$$V = \pi\left[h\left(R^2 - \frac{h^2}{3}\right) - R^2 D \sin^{-1}\left(\frac{h}{R}\right)\right]$$

Surface area excluding base:

$$A = 2\pi R\left(h - D\sin^{-1}\frac{h}{R}\right)$$

Shells

Figure	General properties	Moment of inertia	Radius of gyration, P
Lateral cylindrical shell	Surface area: $2\pi RH$ Centroid: $\bar{z} = \dfrac{H}{2}$	$I_x = I_y = \dfrac{W}{2}\left(R^2 + \dfrac{H^2}{6}\right)$ $I_z = WR^2$ $I_{x_1} = I_{y_1} = \dfrac{W}{6}(3R^2 + 2H^2)$	$P_x = P_y = 0.707\sqrt{R^2 + \dfrac{H}{6}}$ $P_{x_1} = P_{y_1} = 0.408\sqrt{3R^2 + 2H^2}$ $P_z = R$
Total cylindrical shell	Surface area: $2\pi R(R + H)$ Centroid: $\bar{z} = \dfrac{H}{2}$	$I_x = I_y = \dfrac{W}{12(R + H)}$ $\times [3R^2(R + 2H) + H^2(3R + H)]$ $I_{x_1} = I_{y_1} = \dfrac{W}{12(R + H)}$ $\times [3R^2(R + 2H) + 2H^2(3R + H)]$ $I_z = \dfrac{WR^2}{2}\left(\dfrac{R + 2H}{R + H}\right)$	$P_x = P_y = 0.289$ $\times \sqrt{\dfrac{3R^2(R + 2H) + H^2(3R + H)}{R + H}}$ $P_{x_1} = P_{y_1} = 0.289$ $\times \sqrt{\dfrac{3R^2(R + 2H) + 2H^2(3R + H)}{R + H}}$ $P_z = 0.707R\sqrt{\dfrac{R + 2H}{R + H}}$

Shells, continued

Figure	General properties	Moment of inertia	Radius of gyration, P

Total elliptical shell

Surface area:

$\pi H(3A^2 + B^2)/2A$

$$I_x = \frac{WB^2}{4}\left[\frac{7A^2 + B^2}{3A^2 + B^2}\right] + \frac{WH^2}{12}$$

$$I_y = \frac{WA^2}{4}\left[\frac{7B^2 + A^2}{3A^2 + B^2}\right] + \frac{WH^2}{12}$$

$$I_z = \frac{W}{4}\left[\frac{A^4 + 14A^2B^2 + B^4}{3A^2 + B^2}\right]$$

$$I_x = \frac{W}{12}(B^2 + C^2) + \frac{W}{6}\left[\frac{(ABC)(B+C)}{AB + BC + AC}\right]$$

$$I_y = \frac{W}{12}(A^2 + B^2) + \frac{W}{6}\left[\frac{(ABC)(A+B)}{AB + BC + AC}\right]$$

$$I_z = \frac{W}{12}(A^2 + C^2) + \frac{W}{6}\left[\frac{(ABC)(A+C)}{AB + BC + AC}\right]$$

Hollow box

Surface area:

$2(AB + BC + AC)$

Surface area (hollow box with open ends):

$2C(A + B)$

$$I_x = \frac{W}{12}(B^2 + C^2) + \frac{WAB^2}{6(A+B)}$$

$$I_y = \frac{W}{12}(A + B)^2$$

$$I_z = \frac{W}{12}(A^2 + C^2) + \frac{WBA^2}{6(A+B)}$$

Shells, continued

Figure	General properties	Moment of inertia	Radius of gyration, P
Lateral surface of a circular cone 	Surface area: $\pi R \sqrt{R^2 + H^2}$ Centroid: $\bar{z} = \dfrac{H}{3}$	$I_x = I_y = \dfrac{W}{4}\left(R^2 + \dfrac{2}{9}H^2\right)$ $I_z = \dfrac{WR^2}{2}$ $I_{x_1} = I_{y_1} = \dfrac{W}{12}(3R^2 + 2H^2)$	$P_x = P_y = \dfrac{\sqrt{9R^2 + 2H^2}}{6}$ $P_{x_1} = P_{y_1} = 0.289\sqrt{3R^2 + 2H^2}$ $P_z = 0.707R$
Lateral surface of frustum of circular cone 	Surface area: $\pi(R + r)\sqrt{H^2 + (R - r)^2}$ Centroid: $\bar{z} = \dfrac{H}{3}\left(\dfrac{2r + R}{r + R}\right)$	$I_x = I_y = \dfrac{W}{4}(R^2 + r^2)$ $\quad + \dfrac{WH^2}{18}\left[1 + \dfrac{2Rr}{(R+r)^2}\right]$ $I_z = \dfrac{W}{2}(R^2 + r^2)$	$P_x = P_y = \sqrt{\dfrac{(R^2 + r^2)}{4} + \dfrac{H^2}{18}\left[1 + \dfrac{2Rr}{(R+r)^2}\right]}$ $P_z = 0.707\sqrt{R^2 + r^2}$

Shells, continued

Figure	General properties	Moment of inertia	Radius of gyration, P
Spherical shell 	Surface area: $$4\pi R^2$$	$$I_x = I_y = I_z = \frac{2}{3} W R^2$$	$$P_x = P_y = P_z = 0.816R$$
Hemispherical shell 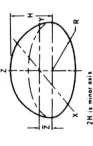	Surface area: $$2\pi R^2$$ Centroid: $$\bar{z} = \frac{R}{2}$$	$$I_x = I_y = \frac{5}{12} W R^2$$ $$I_z = I_{x_1} = I_{y_1} = \frac{2}{3} W R^2$$	$$P_x = P_y = 0.646R$$ $$P_z = P_{x_1} = P_{y_1} = 0.816R$$
Elliptical hemispheroidal shell	Surface area: $$\pi \left[R^2 + \frac{H^2}{2E} \, \mathrm{LOG}_e \, \frac{1+E}{1-E} \right]$$ Centroid: $$\bar{Z} = \frac{2\pi H (R^3 - H^3)}{3E^2 R(\text{Surface Area})}$$ $$E = \frac{\sqrt{R^2 - H^2}}{R(\text{eccentricity})}$$	Notes: Surface area formula is for $\frac{1}{2}$ of an oblate spheroid, i.e., an ellipse rotated about its minor axis 2H. (Looks like M & M candy.) Surface area for a prolate spheroid, i.e., ellipse rotated about its major axis 2R. (Looks like a watermelon.) $2\pi(H^2 + \frac{RH}{E} \sin^{-1} E)$	

2 H is minor axis

Shells, continued

Figure	General properties	Moment of inertia
Paraboloid of revolution shell	Surface area: $\dfrac{\pi R}{6H^2}[P - R^3]$ Centroid: $\bar{Z} = \dfrac{P(6H^2 - R^2) + R^5}{10H(P - R^3)}$ $P = (4H^2 + R^2)^{\frac{3}{2}}$	$I_x = I_y = \dfrac{W}{28H^2}\left[\dfrac{P(12H^4 + 6R^2H^2 - R^4) + R^7}{P - R^3}\right]$ $-\dfrac{W}{100H^2}\left[\dfrac{P(6H^2 - R^2) + R^5}{P - R^3}\right]^2$ $I_z = \dfrac{WR^2}{10H^2}\left[\dfrac{P(6H^2 - R^2) + R^5}{P - R^3}\right]$ or $I_z = \dfrac{WR^2\bar{Z}}{H}$
Sector of a hollow torus	$A = r_0^2 + r_i^2$ $K = \dfrac{2\sin^2 a}{a}\left[2R^2 + A + \dfrac{A^2}{BR^2}\right]$ Length: $2aR$ Centroid: $\bar{X} = \dfrac{\sin a}{a}\left[R + \dfrac{A}{4R}\right]$ Volume: $2\pi Ra(r_0^2 - r_i^2)$	$I_x = \dfrac{W}{16a}[4R^2(2a - \sin 2a) + A(10a - 3\sin 2a)]$ $I_y = \dfrac{W}{16a}[4R^2(2a + \sin 2a) + A(10a + 3\sin 2a) - 4K]$ $I_z = \dfrac{W}{4a}[4R^2a + 3Aa - K]$

Section 3

CONVERSION FACTORS

Length (l)

Item	cm	m	km	in.	ft	mile
1 centimeter (cm)	1	10^{-2}	10^{-5}	0.3937	3.281×10^{-2}	6.214×10^{-6}
1 meter[a] (m)	100	1	10^{-3}	39.37	3.281	6.214×10^{-4}
1 kilometer (km)	10^5	1000	1	3.937×10^4	3281	0.6214
1 inch (in.)	2.540	2.540×10^{-2}	2.540×10^{-5}	1	8.333×10^{-2}	1.578×10^{-5}
1 foot (ft)	30.48	0.3048	3.048×10^{-4}	12	1	1.894×10^{-4}
1 statute mile	1.609×10^5	1609	1.609	6.336×10^4	5280	1

[a]Denotes SI units.

1 foot = 1200/3937 m
1 meter = 3937/1200 ft
1 angstrom (Å) = 10^{-10} m
1 X-unit = 10^{-13} m
1 micron (μ) = 10^{-6} m
1 nanometer = 10^{-9} m

1 light year = 9.460×10^{12} km
1 parsec = 3.084×10^{13} km
1 fathom = 6 ft
1 yard = 3 ft
1 rod = 16.5 ft
1 mil = 10^{-3} in.

1 nautical mile (n mile) = 1852 m = 1.1508 statute miles = 6076.10 ft

Area (A)

Item	m^2	cm^2	ft^2	$in.^2$	circ mil
1 square meter[a] (m^2)	1	10^4	10.76	1550	1.974×10^9
1 square cm	10^{-4}	1	1.076×10^{-3}	0.1550	1.974×10^5
1 square ft	9.290×10^{-2}	929.0	1	144	1.833×10^8
1 square in.	6.452×10^{-4}	6.452	6.944×10^{-3}	1	1.273×10^6
1 circular mil	5.067×10^{-10}	5.067×10^{-6}	5.454×10^{-9}	7.854×10^{-7}	1^b

[a]Denotes SI units.
[b]1 circular mil $= \pi(d)^2/4$, where d is measured in units of 0.001 in. = 1 mil.

1 square mile $= 27,878,400\,ft^2 = 640$ acres; 1 acre $= 43,560\,ft^2$; 1 barn $= 10^{-28}\,m^2$

Volume (V)

Item	m^3	cm^3	l	ft^3	$in.^3$
1 cubic m	1	10^6	1000	35.31	6.102×10^4
1 cubic cm	10^{-6}	1	1.000×10^{-3}	3.531×10^{-5}	6.102×10^{-2}
1 liter[a] (l)	1.000×10^{-3}	1000	1	3.531×10^{-2}	61.02
1 cubic ft	2.832×10^{-2}	2.832×10^4	28.32	1	1728
1 cubic in.	1.639×10^{-5}	16.39	1.639×10^{-2}	5.787×10^{-4}	1

[a]Denotes SI units.

1 U.S. fluid gallon = 4 U.S. fluid quarts = 8 U.S. fluid pints = 128 U.S. fluid ounces $= 231\,in.^3$; 1 British imperial gallon $= 277.42\,in.^3$ (volume of 10 lb H_2O at 62°F); 1 liter $= 1000.028\,cm^3$ (volume of 1 kg H_2O at its maximum density)

Plane Angle (θ)

Item	deg	min	s	rad	rev
1 degree (deg)	1	60	3600	1.745×10^{-2}	2.778×10^{-3}
1 minute (min or ′)	1.667×10^{-2}	1	60	2.909×10^{-4}	4.630×10^{-5}
1 second[a] (s or ″)	2.778×10^{-4}	1.667×10^{-2}	1	4.848×10^{-6}	7.716×10^{-7}
1 radian[a] (rad)	57.30	3438	2.063×10^5	1	0.1592
1 revolution (rev)	360	2.16×10^4	1.296×10^5	6.283	1

[a]Denotes SI units.

1 rev $= 2\pi$ rad $= 360$ deg; 1 deg $= 60' = 3600''$

Solid Angle

1 sphere $= 4\pi$ steradians $= 12.57$ steradians

Mass (m)

Item	g	kg	slug	amu	oz	lb	ton
1 gram (g)	1	0.001	6.852×10^{-5}	6.024×10^{23}	3.527×10^{-2}	2.205×10^{-3}	1.102×10^{-6}
1 kilogram[a] (kg)	1000	1	6.852×10^{-2}	6.024×10^{26}	35.27	2.205	1.102×10^{-3}
1 slug	1.459×10^{4}	14.59	1	8.789×10^{27}	514.8	32.17	1.609×10^{-2}
1 atomic mass unit (amu)	1.600×10^{-24}	1.660×10^{-27}	1.137×10^{-28}	1	5.855×10^{-26}	3.660×10^{-27}	1.829×10^{-30}
1 ounce (avoirdupois)	28.35	2.835×10^{-2}	1.943×10^{-3}	1.708×10^{25}	1	6.250×10^{-2}	3.125×10^{-5}
1 pound (avoirdupois)	453.6	0.4536	3.108×10^{-2}	2.732×10^{26}	16	1	0.0005
1 ton	9.072×10^{5}	907.2	62.16	5.465×10^{29}	3.200×10^{4}	2000	1

[a] Denotes SI units.

Note: Portion of table enclosed in the box must be used with caution because those units are not properly mass units but weight equivalents, which depend on standard terrestrial acceleration due to gravity, g, where $g = 9.80665$ m/s^2 = 32.174 ft/s^2 on Earth at sea level. The conversion between the equivalent weight (lb) and the true mass parameter (lbm = pound mass) is given by 1.0 lb = 32.174 lbm·ft/s^2.

1 metric ton = 1000 kg

Density (ρ)

Item	slug/ft^3	kg/m^3	g/cm^3	lb/ft^3	lb/in.^{3}
1 slug per ft^3	1	515.4	0.5154	32.17	1.862×10^{-2}
1 kg per m^3	1.940×10^{-3}	1	0.001	6.243×10^{-2}	3.613×10^{-5}
1 g per cm^3	1.940	1000	1	62.43	3613×10^{-2}
1 lb per ft^3	3.108×10^{-2}	16.02	1.602×10^{-2}	1	5.787×10^{-4}
1 lb per in.^{3}	53.71	2.768×10^4	27.68	1728	1

Note: Portion of table enclosed in the box must be used with caution because those units are not mass-density units but weight-density units, which depend on g.

Time (t)

Item	yr	day	h	min	s
1 year (yr)	1	365.2^a	8.766×10^3	5.259×10^5	3.156×10^7
1 solar day	2.738×10^{-3}	1	24	1440	8.640×10^4
1 hour (h)	1.41×10^{-4}	4.167×10^{-2}	1	60	3600
1 minute (min)	1.901×10^{-6}	6.944×10^{-4}	1.667×10^{-2}	1	60
1 second^b (s)	3.169×10^{-8}	1.157×10^{-5}	2.778×10^{-4}	1.667×10^{-2}	1
1 day (sidereal)	2.73×10^{-3}	0.99726	23.93447	1436.068	86,164.091

^{a}1 year = 365.24219879 days.
^{b}Denotes SI units.

Velocity (v)

Item	ft/s	km/h	m/s	mile/h	cm/s	kn
1 ft per s	1	1.097	0.3048	0.6818	30.48	0.5925
1 km per h	0.9113	1	0.2778	0.6214	27.78	0.5400
1 m per s^a	3.281	3.600	1	2.237	100	1.944
1 mile per h	1.467	1.609	0.4470	1	44.70	0.8689
1 cm per s	3.281×10^{-2}	3.600×10^{-2}	0.0100	2.237×10^{-2}	1	1.944×10^{-2}
1 knot (kn)	1.688	1.852	0.5144	1.151	51.44	1

^{a}Denotes SI units.

1 knot = 1 n mile/h; 1 mile/min = 88 ft/s = 60 mile/h
Speed of light in a vacuum = 2.99792458×10^8 m/s

Force (F)

Item	dyne	N	lb	poundal	gf	kgf
1 dyne	1	10^{-5}	2.248×10^{-6}	7.233×10^{-5}	1.020×10^{-3}	1.020×10^{-6}
1 newton (N)	10^5	1	0.2248	7.233	102.0	0.1020
1 pound (lb)	4.448×10^5	4.448	1	32.174	453.6	0.4536
1 poundal	1.383×10^4	0.1383	3.108×10^{-2}	1	14.10	1.410×10^{-2}
1 gram-force (gf)	980.7	9.807×10^{-3}	2.205×10^{-3}	7.093×10^{-2}	1	0.001
1 kilogram-force (kgf)	9.807×10^5	9.807	2.205	70.93	1000	1

Note: Portion of table enclosed in the box must be used with caution because those units are not force units but weight equivalents of mass, which depend on g.

$$1 \text{ kgf} = 9.80665 \text{ N}; \quad 1 \text{ lb} = 32.174 \text{ poundal}$$

Pressure (p)

Item	atm	dyne/cm²	in. H$_2$O	cm Hg	N/m² (Pa)	lb/in.²	lb/ft²
1 atmosphere (atm)	1	1.013×10^6	406.8	76	1.013×10^5	14.70	2116
1 dyne per cm²	9.869×10^{-7}	1	4.015×10^{-4}	7.501×10^{-5}	0.100	1.450×10^{-5}	2.089×10^{-3}
1 in. of water at 4°C[a]	2.458×10^{-3}	2491	1	0.1868	249.1	3.613×10^{-2}	5.202
1 cm of mercury at 0°C[a]	1.316×10^{-2}	1.333×10^4	5.353	1	1333	0.1934	27.85
1 N per m²	9.869×10^{-6}	10	4.015×10^{-3}	7.501×10^{-4}	1	1.450×10^{-4}	2.089×10^{-2}
1 lb per in.²	6.805×10^{-2}	6.895×10^4	27.68	5.171	6.895×10^3	1	144
1 lb per ft²	4.725×10^{-4}	478.8	0.1922	3.591×10^{-2}	47.88	6.944×10^{-3}	1

[a]Where the acceleration of gravity has the standard value 9.80665 m/s².

$$1 \text{ bar} = 10^6 \text{ dyne/cm}^2 = 1 \text{ kPa}$$

Energy, Work, Heat (W)

Item	Btu	erg	ft-lb	hp-h	J
1 British thermal unit (Btu)	1	1.055×10^{10}	777.9	3.929×10^{-4}	1055
1 erg	9.481×10^{-11}	1	7.376×10^{-8}	3.725×10^{-14}	10^{-7}
1 foot-pound (ft-lb)	1.285×10^{-3}	1.356×10^{7}	1	5.051×10^{-7}	1.356
1 horsepower-hour (hp-h)	2545	2.685×10^{13}	1.980×10^{6}	1	2.685×10^{6}
1 joule[a] (J)	9.481×10^{-4}	10^{7}	0.7376	3.725×10^{-7}	1
1 calorie (cal)	3.968×10^{-3}	4.186×10^{7}	3.087	1.559×10^{-6}	4.187
1 kilowatt-hour (kW-h)	3413	3.6×10^{13}	2.655×10^{6}	1.341	3.6×10^{6}
1 electron volt (eV)	1.519×10^{-22}	1.602×10^{-12}	1.182×10^{-19}	5.967×10^{-26}	1.602×10^{-19}
1 million electron volts	1.519×10^{-16}	1.602×10^{-6}	1.182×10^{-13}	5.967×10^{-20}	1.602×10^{-13}
1 kg	8.521×10^{13}	8.987×10^{23}	6.629×10^{16}	3.348×10^{10}	8.987×10^{16}
1 amu	1.415×10^{-13}	1.492×10^{-3}	1.100×10^{-10}	5.558×10^{-17}	1.492×10^{-10}

[a]Denotes SI units.

Note: The electron volt is the kinetic energy an electron gains from being accelerated through the potential difference of one volt in an electric field. The units enclosed in the box are not properly energy units; they arise from the relativistic mass-energy equivalent formula, $E = mc^2$.

1 m-kgf = 9.807 J; 1 W-s = 1 J = 1 N-m, 1 dyne-cm = 1 erg

(continued)

Energy, Work, Heat (W), continued

Item	cal	kW-h	eV	MeV	kg	amu
1 British thermal unit (Btu)	252.0	2.930×10^{-4}	6.585×10^{21}	6.585×10^{15}	1.174×10^{-14}	7.074×10^{12}
1 erg	2.389×10^{-8}	2.778×10^{-14}	6.242×10^{11}	6.242×10^{5}	1.113×10^{-24}	670.5
1 foot-pound (ft-lb)	0.3239	3.766×10^{-7}	8.464×10^{18}	8.464×10^{12}	1.509×10^{-17}	9.092×10^{9}
1 horsepower-hour (hp-h)	6.414×10^{5}	0.7457	1.676×10^{25}	1.676×10^{14}	2.988×10^{-11}	1.800×10^{16}
1 joule[a] (J)	0.2389	2.778×10^{-7}	6.242×10^{18}	6.242×10^{12}	1.113×10^{-17}	6.705×10^{7}
1 calorie (cal)	1	1.163×10^{-6}	2.613×10^{19}	2.613×10^{13}	4.659×10^{-17}	2.807×10^{10}
1 kilowatt-hour (kW-h)	8.601×10^{5}	1	2.247×10^{25}	2.247×10^{19}	4.007×10^{-11}	2.414×10^{16}
1 electron volt (eV)	3.827×10^{-20}	4.450×10^{-26}	1	10^{-6}	1.783×10^{-36}	1.074×10^{-9}
1 million electron volts	3.827×10^{-14}	4.450×10^{-20}	10^{6}	1	1.783×10^{-30}	1.074×10^{-3}
1 kg	2.147×10^{16}	2.497×10^{10}	5.610×10^{35}	5.610×10^{29}	1	6.025×10^{26}
1 amu	3.564×10^{-11}	4.145×10^{-17}	9.310×10^{8}	931.0	1.660×10^{-27}	1

[a]Denotes SI units.

Notes: The electron volt is the kinetic energy an electron gains from being accelerated through the potential difference of one volt in an electric field. The units enclosed in the box are not properly energy units; they arise from the relativistic mass-energy equivalent formula, $E = mc^2$.

Power (P)

Item	Btu/h	Btu/s	ft-lb/min	ft-lb/s	hp	cal/s	kW	W
1 Btu/h	1	2.778×10^{-4}	12.97	0.2161	3.929×10^{-4}	7.000×10^{-2}	2.930×10^{-4}	0.2930
1 Btu/s	3600	1	4.669×10^4	777.9	1.414	252.0	1.055	1.055×10^{-3}
1 ft-lb/min	7.713×10^{-2}	2.142×10^{-5}	1	1.667×10^{-2}	3.030×10^{-5}	5.399×10^{-3}	2.260×10^{-5}	2.260×10^{-2}
1 ft-lb/s	4.628	1.286×10^{-3}	60	1	1.818×10^{-3}	0.3239	1.356×10^{-3}	1.356
1 hp	2545	0.7069	3.3×10^4	550	1	178.2	0.7457	745.7
1 cal/s	14.29	0.3950	1.852×10^2	3.087	5.613×10^{-3}	1	4.186×10^{-3}	4.186
1 kW	3413	0.9481	4.425×10^4	737.6	1.341	238.9	1	1000
1 W	3.413	9.481×10^{-4}	44.25	0.7376	1.341×10^{-3}	0.2389	0.001	1

1 W = 1 J/s

Thermal Conductivity (k)

Item	cal/s-cm °C	W/m K	W/in. °C	Btu/h-ft °F	Btu/s-in. °F	hp/ft °F
1 cal per s per cm per °C	1	418.5	10.63	241.9	5.600×10^{-3}	9.503×10^{-2}
1 W per m per K[a]	2.390×10^{-3}	1	2.540×10^{-2}	0.5781	1.338×10^{-5}	2.271×10^{-4}
1 W per in. per °C	9.407×10^{-2}	39.37	1	22.76	5.269×10^{-4}	8.939×10^{-3}
1 Btu per h per ft per °F	4.134×10^{-3}	1.730	4.394×10^{-2}	1	2.315×10^{-5}	3.929×10^{-4}
1 Btu per s per in. per °F	1.786×10^{2}	7.474×10^{4}	1.898×10^{3}	4.320×10^{4}	1	16.97

[a] Denotes SI units.

Absolute or Dynamic Viscosity (μ)

Item	centipoise	poise	kgf-s/m²	lb-s/ft²	kg/m-s	lbm/ft-s
1 centipoise	1	10^{-2}	1.020×10^{-4}	2.089×10^{-5}	10^{-3}	6.720×10^{-4}
1 poise	100	1	1.020×10^{-2}	2.089×10^{-3}	0.100	6.720×10^{-2}
1 N-s per m²	9.807×10^{3}	98.07	1	0.2048	9.807	6.590
1 lb (force)-s per ft²	4.788×10^{4}	4.788×10^{2}	4.882	1	47.88	32.174
1 kg per m-s	10^{3}	10	0.1020	2.089×10^{-2}	1	0.6720
1 lb (mass) per ft-s	1.488×10^{3}	14.88	0.1518	3.108×10^{-2}	1.488	1

Note: The absolute viscosity μ is properly expressed in force units according to its definition. In heat transfer and fluid mechanics it is usually expressed in mass-equivalent units to avoid the use of a conversion factor in Reynolds number. Mass equivalent units have been used in the portion of the table enclosed in the box. The proper force unit for μ in the SI system is N-s per m²; it is seldom used. The poise, the cgs absolute viscosity unit, is defined as 1 dyne-s/cm².

Inductance (L)

Item	abhenry	henry	μH	stathenry
1 abhenry[a] (1 emu)	1	10^{-9}	0.001	1.113×10^{-21}
1 henry[b] (H)	10^9	1	10^6	1.113×10^{-12}
1 microhenry (μH)	10^3	10^{-6}	1	1.113×10^{-18}
1 stathenry[a] (1 esu)	8.987×10^{20}	8.987×10^{11}	8.987×10^{17}	1

[a]These units are listed for historical completeness only. They are no longer used.
[b]Denotes SI units.

Capacitance (C)

Item	abF	F	μF	staff
1 abfarad[a] (1 emu)	1	10^9	10^{15}	8.987×10^{20}
1 farad[b] (F)	10^{-9}	1	10^6	8.987×10^{11}
1 microfarad (μF)	10^{-15}	10^{-6}	1	8.987×10^5
1 statfarad[a] (1 esu)	1.113×10^{-21}	1.113×10^{-12}	1.113×10^{-6}	1

[a]These units are listed for historical completeness only. They are no longer used.
[b]Denotes SI units.

Kinematic Viscosity ($v = \mu/\rho$)

Item	centistoke	stoke	m^2/s	ft^2/s
1 centistoke	1	10^{-2}	10^{-6}	1.076×10^{-5}
1 stoke	100	1	10^{-4}	1.076×10^{-3}
1 m^2/s[a]	10^6	10^4	1	10.76
1 ft^2/s	9.290×10^4	929.0	9.290×10^{-2}	1

[a]Denotes SI units.

1 stoke = 1 cm^2/s

Electrical Resistance (R)

Item	abohm	ohm	statohm
1 abohm[a] (1 emu)	1	10^{-9}	1.113×10^{-21}
1 ohm[b]	10^9	1	1.113×10^{-12}
1 statohm[a] (1 esu)	8.987×10^{20}	8.987×10^{11}	1

[a]These units are listed for historical completeness only. They are no longer used.
[b]Denotes SI units.

Electrical Resistivity, Reciprocal Conductivity (ρ)

Item	abohm-cm	ohm-cm	ohm-m	statohm-cm	ohm-circ mil/ft[a]
1 abohm-cm (1 emu)	1	10^{-9}	10^{-11}	1.113×10^{-21}	6.015×10^{-3}
1 ohm-cm	10^9	1	0.0100	1.113×10^{-12}	6.015×10^6
1 ohm-m[b]	10^{11}	100	1	1.113×10^{-10}	6.015×10^8
1 statohm-cm (1 esu)	8.987×10^{20}	8.987×10^{11}	8.987×10^9	1	5.406×10^{18}
1 ohm-circular mil per ft	166.2	1.662×10^{-7}	1.662×10^{-9}	1.850×10^{-19}	1

[a] 1 circular mil = $\pi(d)^2/4$, where d is measured in units of 0.001 in. = 1 mil.
[b] Denotes SI units.

Magnetic Field Intensity (H)

Item	abamp-turn/cm	amp-turn/cm	amp-turn/m	amp-turn/in.	oersted
1 abamp-turn per cm	1	10	1000	25.40	12.57
1 amp-turn per cm	0.100	1	100	2.54	1.257
1 amp-turn per m[a]	10^{-3}	10^{-2}	1	2.540×10^{-2}	1.257×10^{-2}
1 amp-turn per in.	3.937×10^{-2}	0.3937	39.37	1	0.4947
1 oersted	7.958×10^{-2}	0.7958	79.58	2.021	1

[a]Denotes SI units.

1 oersted = 1 gilbert/cm; 1 esu = 2.655×10^{-9} amp-turn/m; 1 praoersted = 4π amp-turn/m

Magnetomotive Force

Item	abamp-turn	amp-turn	gilbert
1 abamp-turn	1	10	12.57
1 amp-turn[a]	0.100	1	1.257
1 gilbert	7.958×10^{-2}	0.7958	1

[a]Denotes SI units.

1 pragilbert = 4π amp-turn
1 esu = 2.655×10^{-11} amp-turn

Magnetic Flux (ϕ)

Item	maxwell	kiloline	wb
1 maxwell (1 line or 1 emu)	1	0.001	10^{-8}
1 kiloline	1000	1	10^{-5}
1 weber[a] (wb)	10^8	10^5	1

[a]Denotes SI units.

1 esu = 2.998 wb

Magnetic Flux Density (B)

Item	gauss	kiloline/in.^{2}	wb/m^2	milligauss	gamma
1 gauss (line per cm^2)	1	6.452×10^{-3}	10^{-4}	1000	10^5
1 kiloline per in.^{2}	155.0	1	1.550×10^{-2}	1.550×10^5	1.550×10^7
1 wb per m^{2}[a] (T)	10^4	64.52	1^b	10^7	10^9
1 milligauss	10^{-3}	6.452×10^{-6}	10^{-7}	1	100
1 gamma	10^{-5}	6.452×10^{-8}	10^{-9}	10^{-2}	1

[a]Denotes SI units.
[b]1 wb/m^2 = 1 tesla (T).

1 esu = 2.998×10^6 wb/m^2

Machine Screws—Tap Drill and Clearance Drill Sizes

Screw size		Threads per in.		Drill sizes	
No.	O.D.	N.C.	N.F.	Tap[a]	Clear
0	0.060		80	3/64	#51
1	0.073	64		#53	#47
1	0.073		72	#53	#47
2	0.086	56		#50	#42
2	0.086		64	#50	#42
3	0.099	48		#47	#37
3	0.099		56	#46	#37
4	0.112	40		#43	#31
4	0.112		48	3/32	#31
5	0.125	40		#38	#29
5	0.125		44	#37	#29
6	0.138	32		#36	#27
6	0.138		40	#33	#27
8	0.164	32		#29	#18
8	0.164		36	#29	#18
10	0.190	24		#26	#9
10	0.190		32	#21	#9
12	0.216	24		#16	#2
12	0.216		28	#15	#2
		Fractional sizes start here			
1/4	0.250	20		#7	17/64
1/4	0.250		28	#3	17/64
5/16	0.312	18		F	21/64
5/16	0.312		24	I	21/64
3/8	0.375	16		5/16	25/64
3/8	0.375		24	Q	25/64
7/16	0.437	14		U	29/64
7/16	0.437		20	25/64	29/64
1/2	0.500	13		27/64	33/64
1/2	0.500		20	29/64	33/64
9/16	0.562	12		31/64	37/64
9/16	0.562		18	33/64	37/64
5/8	0.625	11		17/32	41/64
5/8	0.625		18	37/64	41/64
3/4	0.750	10		21/32	49/64
3/4	0.750		16	11/16	49/64
7/8	0.875	9		49/64	57/64
7/8	0.875		14	13/16	57/64
1	1.000	8		7/8	1-1/64
1	1.000		14	15/16	1-1/64

[a]Tap drill sizes shown give approximately 75% depth of thread.

Decimal Equivalents of Drill Size

Letter sizes		Number sizes			
Letter	Sizes in in.	No.	Sizes in in.	No.	Sizes in in.
A	0.234	1	0.2280	41	0.0960
B	0.238	2	0.2210	42	0.0935
C	0.242	3	0.2130	43	0.0890
D	0.246	4	0.2090	44	0.0860
E	0.250	5	0.2055	45	0.0820
F	0.257	6	0.2040	46	0.0810
G	0.261	7	0.2010	47	0.0785
H	0.266	8	0.1990	48	0.0760
I	0.272	9	0.1960	49	0.0730
J	0.277	10	0.1935	50	0.0700
K	0.281	11	0.1910	51	0.0670
L	0.290	12	0.1890	52	0.0635
M	0.295	13	0.1850	53	0.0595
N	0.302	14	0.1820	54	0.0550
O	0.316	15	0.1800	55	0.0520
P	0.323	16	0.1770	56	0.0465
Q	0.332	17	0.1730	57	0.0430
R	0.339	18	0.1695	58	0.0420
S	0.348	19	0.1660	59	0.0410
T	0.358	20	0.1610	60	0.0400
U	0.368	21	0.1590	61	0.0390
V	0.377	22	0.1570	62	0.0380
W	0.386	23	0.1540	63	0.0370
X	0.397	24	0.1520	64	0.0360
Y	0.404	25	0.1495	65	0.0350
Z	0.413	26	0.1470	66	0.0330
		27	0.1440	67	0.0320
		28	0.1405	68	0.0310
		29	0.1360	69	0.0292
		30	0.1285	70	0.0280
		31	0.1200	71	0.0260
		32	0.1160	72	0.0250
		33	0.1130	73	0.0240
		34	0.1110	74	0.0225
		35	0.1100	75	0.0210
		36	0.1065	76	0.0200
		37	0.1040	77	0.0180
		38	0.1015	78	0.0160
		39	0.0995	79	0.0145
		40	0.0980	80	0.0135

Standard Gages

This table shows the standard gages and the names of major commodities for which each is used. To determine the gage used for any commodity, note the number in parentheses opposite the commodity named and find the gage column below bearing the same number in parentheses.

Ga. No.	(1) Birmingham or Stubs	(2) American or Browne & Sharpe	(3) U.S. Standard	(4) Washburn & Moen	(5) Music wire (std.)	(6) Mfgrs. std. ga. for sheet metal
1/0	0.340	0.3249	0.3125	0.3065	0.009	—
1	0.300	0.2893	0.2812	0.2830	0.010	—
2	0.284	0.2576	0.2656	0.2625	0.011	
3	0.259	0.2294	0.2500	0.2437	0.012	0.2391
4	0.238	0.2043	0.2343	0.2253	0.013	0.2242
5	0.220	0.1819	0.2187	0.2070	0.014	0.2092
6	0.203	0.1620	0.2031	0.1920	0.016	0.1943
7	0.180	0.1443	0.1875	0.1770	0.018	0.1793
8	0.165	0.1285	0.1718	0.1620	0.020	0.1644
9	0.148	0.1144	0.1562	0.1483	0.022	0.1495
10	0.134	0.1019	0.1406	0.1350	0.024	0.1345
11	0.120	0.0907	0.1250	0.1205	0.026	0.1196
12	0.109	0.0808	0.1093	0.1055	0.029	0.1046
13	0.095	0.0719	0.0937	0.0915	0.031	0.0897
14	0.083	0.0640	0.0781	0.0800	0.033	0.0747
15	0.072	0.0570	0.0703	0.0720	0.035	0.0673
16	0.065	0.0508	0.0625	0.0625	0.037	0.0598
17	0.058	0.0452	0.0562	0.0540	0.039	0.0538

Commodity:

Aluminum (2) except tubing (1)
Bands (1)
Brass tubing (3/8" O.D. and larger) (1)
Brass tubing (smaller than 3/8" O.D.) (2)
Brass sheets (2)
Brass strips (2)
Brass wire (2)
Copper sheets (2)
Copper wire (2)
Flat wire (1)
Hoops (1)
Iron wire (4)
Monel metal sheets (3)
Music wire (5)
Nickel sheets (3)
Nickel silver sheets (2)
Nickel silver wire (2)

(continued)

Standard Gages, continued

Ga. No.	(1) Birmingham or Stubs	(2) American or Browne & Sharpe	(3) U.S. Standard	(4) Washburn & Moen	(5) Music wire (std.)	(6) Mfgrs. std. ga. for sheet metal
18	0.049	0.0403	0.0500	0.0475	0.041	0.0478
19	0.042	0.0359	0.0437	0.0410	0.043	0.0418
20	0.035	0.0319	0.0375	0.0348	0.045	0.0359
21	0.032	0.0284	0.0343	0.0317	0.047	0.0329
22	0.028	0.0253	0.0312	0.0286	0.049	0.0299
23	0.025	0.0225	0.0281	0.0258	0.051	0.0269
24	0.022	0.0201	0.0250	0.0230	0.055	0.0239
25	0.020	0.0179	0.0218	0.0204	0.059	0.0209
26	0.018	0.0159	0.0187	0.0181	0.063	0.0179
27	0.016	0.0142	0.0171	0.0173	0.067	0.0164
28	0.014	0.0126	0.0156	0.0162	0.071	0.0149
29	0.013	0.0112	0.0140	0.0150	0.075	0.0135
30	0.012	0.0100	0.0125	0.0140	0.080	0.0120
31	0.010	0.0089	0.0109	0.0132	0.085	0.0105
32	0.009	0.0079	0.0101	0.0128	0.090	0.0097
33	0.008	0.0071	0.0093	0.0118	0.095	0.0090
34	0.007	0.0063	0.0085	0.0104	0.100	0.0082
35	0.005	0.0056	0.0078	0.0095	0.106	0.0075
36	0.004	0.0050	0.0070	0.0090	0.112	0.0067
37		0.0044	0.0066	0.0085	0.118	0.0064
38		0.0039	0.0062	0.0080	0.124	0.0060

Commodity:
Phosphor bronze strip (2)
Spring steel (1)
Stainless steel (3)
Steel plates (6)
Steel sheets (6)
Steel tubing, seamless and welded (1)
Steel wire (4) exceptions:
Music wire (5)
Armature binding wire (2)
Flat wire (1)
Strip steel (1)

Fundamental Constants

Quantity	Symbol	Value
Speed of light in vacuum	c	299,792,458 m/s
Newtonian constant of gravitation	G	6.67259×10^{-11} m^3 kg^{-1} s^{-2}
Planck constant	h	$6.6260755 \times 10^{-34}$ J-s
Elementary charge	e	$1.60217733 \times 10^{-19}$ C
Magnetic flux quantum, $h/2e$	Φ_o	$2.06783461 \times 10^{-13}$ wb
Josephson frequency–voltage ratio	$2e/h$	4.8359767×10^{14} Hz/V
Quantized Hall resistance, h/e^2	R_h	25,812.8056 Ω
Rydberg constant, $m_o c \alpha^2 / 2h$	R_∞	10,973,731.534 M^{-1}
Bohr radius, $\alpha/4\pi R_\infty$	a_o	$0.529177249 \times 10^{-10}$ m
Compton wavelength, $h/m_o c$	λ_c	$2.42631058 \times 10^{-12}$ m
Classical electron radius, $\alpha^2 a_o$	r_e	$2.81794092 \times 10^{-15}$ m
Thompson cross section, $(8\pi/3)r_o^2$	σ_o	$0.66524616 \times 10^{-28}$ m^2
Permeability of vacuum	μ_o	$12.566370614 \times 10^{-7}$ NA^{-2}
Permittivity of vacuum, $1/\mu_o c^2$	ε_o	$8.854187817 \times 10^{-12}$ F/m
Bohr magneton, $eh/2m_o$	μ_B	$9.2740154 \times 10^{-24}$ J/T
Nuclear magneton, $eh/2m_p$	μ_N	$5.0507866 \times 10^{-27}$ J/T
Electron mass	m_e	$9.1093897 \times 10^{-31}$ kg
Electron magnetic moment	μ_e	$928.47701 \times 10^{-26}$ J/T
in Bohr magnetons	μ_e/μ_B	1.001159652193
Muon mass	m_μ	$1.8835327 \times 10^{-28}$ kg
Muon–electron mass ratio	m_μ/m_e	206.768262
Muon magnetic moment	μ_μ	$4.4904514 \times 10^{-26}$ J/T
Muon magnetic moment anomaly, $[\mu_\mu/(eh/2m_\mu)] - 1$	a_μ	0.0011659230
Muon–proton magnetic moment ratio	μ_μ/μ_p	3.18334547
Proton mass	m_p	$1.6726231 \times 10^{-27}$ kg
Proton–electron mass ratio	m_p/m_e	1,836.152701
Proton magnetic moment	μ_p	1.410607×10^{-26} J/T
in Bohr magnetons	μ_p/μ_B	1.521032202
in nuclear magnetons	μ_p/μ_N	2.792847386

(continued)

Fundamental Constants, continued

Quantity	Symbol	Value
Neutron mass	m_n	$1.6749286 \times 10^{-27}$ kg
Neutron–electron mass ratio	m_n/m_e	1,838.683662
Neutron–proton mass ratio	m_n/m_p	1.001378404
Neutron magnetic moment	μ_n	$0.96623707 \times 10^{-26}$ J/T
in nuclear magnetrons	μ_n/μ_N	1.91304275
Neutron–proton magnetic	μ_n/μ_p	0.68497934
moment ratio		
Deutron mass	m_d	$3.3435860 \times 10^{-27}$ kg
Deutron–electron mass ratio	m_d/m_e	3,670.483014
Deutron–proton mass ratio	m_d/m_p	1.999007496
Deutron–proton magnetic	μ_d/μ_p	0.3070122035
moment ratio		
Avogadro constant	N_A, L	6.0221367×10^{23} mol^{-1}
Faraday constant, $N_A e$	F	96,485.309 C/mol
Planck constant, molar	$N_A h$	$3.99031323 \times 10^{-10}$ J-s/mol
	$N_A hc$	0.11962658 Jm/mol
Molar gas constant	R	8.314510 J/mol/K
Boltzmann constant, R/N_A	k	1.380658×10^{-23} J/K
Molar volume (ideal gas), RT/p	V_m	0.02241410 m^3/mol
$T = 273.15$ K, $p = 101,326$ Pa		
Stefan–Boltzmann constant	σ	5.67051×10^{-8} W/m^{2}K^4
First radiation constant, $2\pi hc^2$	c_1	$3.7417749 \times 10^{-16}$ Wm^2
Second radiation constant, hc/k	c_2	0.01438769 mK
Electron mass in atomic		
mass units		5.485802×10^{-4} u
Acceleration of gravity	g	9.80665 m/s^2
		32.17405 ft/s^2
Absolute zero		$-459.688°$F
		$-273.16°$C
Fine structure constant	α	$7.29735308 \times 10^{-3}$

Common Derived Units

Quantity	Unit	Symbol
Acceleration	meter per second squared, m s^{-2}	a
	feet per second squared, ft s^{-2}	a
Acceleration, angular	radian per second squared, rad s^{-2}	α
Activity of a radionuclide	becquerel, (one per second) s^{-1}	Bq
Angular velocity	radian per second, rad s^{-1}	ω
Area	square meter, m^2	A
	square feet, ft^2	A
Capacitance	farad, (coulomb per volt) C V^{-1}	T
Concentration	mole per cubic meter, mol m^{-3}	
	mole per cubic foot, mol ft^{-3}	
Density	kilogram per cubic meter, kg m^{-3}	ρ
	pound per cubic inch, lb in.$^{-3}$	ρ
Dose, absorbed	gray, (joule per kilogram) J kg^{-1}	Gy
Dose, equivalent	sievert, (joule per kilogram) J kg^{-1}	Sv
Electric conductance	siemens, (one per ohm) Ω^{-1}	S
Electric field strength	volt per meter, V m^{-1}	
	volt per foot, V ft^{-1}	
Electric resistance	ohm, (volt per ampere) V amp^{-1}	Ω
Energy, work, quantity of heat	joule, (newton meter) N-m	J
	British thermal unit	Btu
	foot pound, ft-lb	
Force	newton, (meter kilogram per second squared) m kg s^{-2}	N
	pound, lb	
Frequency	hertz, (one per second) s^{-1}	Hz
Heat capacity, entropy	joule per degree Kelvin, J K^{-1}	
	Btu per degree Fahrenheit, Btu °F^{-1}	
Heat flux density, irradiance	watt per square meter, W m^{-2}	
	watt per square foot, W ft^{-2}	
Illuminance	lux, (lumen per square meter) lm m^{-2}	lx
	lux, (lumen per square foot) lm ft^{-2}	lx
Inductance	henry, (ohm second) Ω s	H

(continued)

Common Derived Units, continued

Quantity	Unit	Symbol
Luminance	candela per square meter, cd m^{-2}	
	candela per square foot, cd ft^{-2}	
Luminance flux	luman, cd sr	lm
Magnetic field strength	ampere per meter, amp m^{-1}	
	ampere per foot, amp ft^{-1}	
Magnetic flux	weber, (volt second) V s	wb
Magnetic flux density	tesla, (weber per square meter) wb m^{-2}	T
	tesla, (weber per square foot) wb ft^{-2}	T
Magnetomotive force	ampere	amp
Molar energy	joule per mole, J mol^{-1}	
	Btu per mole, Btu mol^{-1}	
Molar entropy, molar heat capacity	joule per mole Kelvin, J mol^{-1} K^{-1}	
	Btu per mole Fahrenheit, Btu mol^{-1} °F^{-1}	
Potential difference, electromotive force	volt, (watt per ampere) W amp^{-1}	V
Power, radiant flux	watt, (joule per second) J s^{-1}	W
	watt, (Btu per second) Btu s^{-1}	W
Pressure, stress	pascal, (newton per square meter) N m^{-2}	Pa
	atmosphere, (pounds per square inch) lb in.$^{-2}$	P
Quantity of electricity	coulomb, (ampere second) amp s	C
Specific heat capacity, specific entropy	joule per kilogram Kelvin, J kg^{-1} K^{-1}	
	Btu per pound Fahrenheit, Btu lb^{-1} °F^{-1}	
Thermal conductivity	watt per meter Kelvin, W m^{-1} K^{-1}	
	watt per foot Fahrenheit, W ft^{-1} °F^{-1}	
Velocity	meter per second, m s^{-1}	v
	feet per second, ft s^{-1}	v
	miles per hour, mph	v
Viscosity, dynamic	pascal second, Pa s	
	centipoise	
Viscosity, kinematic	meter squared per second, m^2 s^{-1}	
	foot squared per second, ft^2 s^{-1}	
Volume	cubic meter, m^3	
	cubic inch, in.^{3}	
Wave number	one per meter, m^{-1}	
	one per inch, in.$^{-1}$	

Section 4

STRUCTURAL ELEMENTS

Beam Formulas

Bending moment, vertical shear, and deflection of beams of uniform cross section under various conditions of loading

Nomenclature

E = modulus of elasticity, lb/in.^{2}
I = moment of inertia, in.^{4}
l = length of beam, in.
M = maximum bending moment, lb-in.
M_x = bending moment at any section, lb-in.
P = concentrated loads, lb
R_1, R_2 = reactions, lb
V = maximum vertical shear, lb
V_x = vertical shear at any section, lb
w = uniform load per unit of length, lb/in.
W = total uniform load on beam, lb
x = distance from support to any section, in.
y = maximum deflection, in.

Simple Beam—Uniform Load

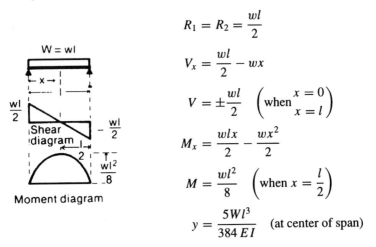

$$R_1 = R_2 = \frac{wl}{2}$$

$$V_x = \frac{wl}{2} - wx$$

$$V = \pm\frac{wl}{2} \quad \left(\text{when}\begin{array}{l} x = 0 \\ x = l \end{array}\right)$$

$$M_x = \frac{wlx}{2} - \frac{wx^2}{2}$$

$$M = \frac{wl^2}{8} \quad \left(\text{when } x = \frac{l}{2}\right)$$

$$y = \frac{5Wl^3}{384\,EI} \quad \text{(at center of span)}$$

The beam formulas appearing on pages 4-2–4-10 are from *Handbook of Engineering Fundamentals*, 3rd Edition, by O. W. Esbach and M. Souders. Copyright © 1976, John Wiley & Sons, Inc., New York. Reprinted by permission of John Wiley & Sons, Inc.

Beam Formulas, continued

Simple Beam—Concentrated Load at Any Point

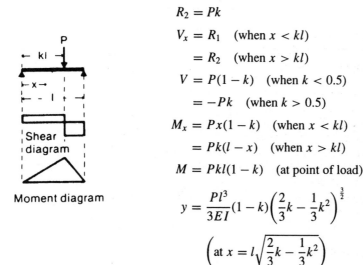

Shear diagram

Moment diagram

$$R_1 = P(1 - k)$$

$$R_2 = Pk$$

$$V_x = R_1 \quad \text{(when } x < kl)$$

$$\quad = R_2 \quad \text{(when } x > kl)$$

$$V = P(1 - k) \quad \text{(when } k < 0.5)$$

$$\quad = -Pk \quad \text{(when } k > 0.5)$$

$$M_x = Px(1 - k) \quad \text{(when } x < kl)$$

$$\quad = Pk(l - x) \quad \text{(when } x > kl)$$

$$M = Pkl(1 - k) \quad \text{(at point of load)}$$

$$y = \frac{Pl^3}{3EI}(1 - k)\left(\frac{2}{3}k - \frac{1}{3}k^2\right)^{\frac{3}{2}}$$

$$\left(\text{at } x = l\sqrt{\frac{2}{3}k - \frac{1}{3}k^2}\right)$$

Simple Beam—Concentrated Load at Center

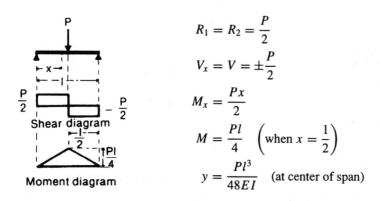

Shear diagram

Moment diagram

$$R_1 = R_2 = \frac{P}{2}$$

$$V_x = V = \pm\frac{P}{2}$$

$$M_x = \frac{Px}{2}$$

$$M = \frac{Pl}{4} \quad \left(\text{when } x = \frac{1}{2}\right)$$

$$y = \frac{Pl^3}{48EI} \quad \text{(at center of span)}$$

Beam Formulas, continued

Simple Beam—Two Equal Concentrated Loads at Equal Distances from Supports

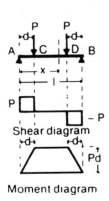

Shear diagram

Moment diagram

$$R_1 = R_2 = P$$

$$V_x = P \qquad \text{for } AC$$
$$= 0 \qquad \text{for } CD$$
$$= -P \qquad \text{for } DB$$
$$V = \pm P$$
$$M_x = Px \qquad \text{for } AC$$
$$= Pd \qquad \text{for } CD$$
$$= P(l - x) \quad \text{for } DB$$
$$M = Pd$$
$$y = \frac{Pd}{24EI}(3l^2 - 4d^2)$$
$$\text{(at center of span)}$$

Simple Beam—Load Increasing Uniformly from Supports to Center of Span

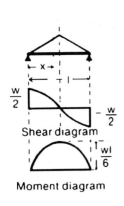

Shear diagram

Moment diagram

$$R_1 = R_2 = \frac{W}{2}$$

$$V_x = W\left(\frac{1}{2} - \frac{2x^2}{l^2}\right) \quad \left(\text{when } x < \frac{l}{2}\right)$$

$$V = \pm\frac{W}{2} \quad \text{(at supports)}$$

$$M_x = Wx\left(\frac{1}{2} - \frac{2x^2}{3l^2}\right)$$

$$M = \frac{Wl}{6} \quad \text{(at center of span)}$$

$$y = \frac{Wl^3}{60EI} \quad \text{(at center of span)}$$

Beam Formulas, continued

Cantilever Beam—Load Concentrated at Free End

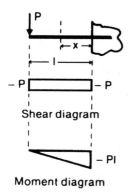

Shear diagram

Moment diagram

$$R = P$$

$$V_x = V = -P$$

$$M_x = -P(l - x)$$

$$M = -Pl \quad \text{(when } x = 0)$$

$$y = \frac{Pl^3}{3EI}$$

Simple Beam—Load Increasing Uniformly from Center to Supports

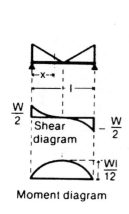

Shear diagram

Moment diagram

$$R_1 = R_2 = \frac{W}{2}$$

$$V_x = -W\left(\frac{2x}{l} - \frac{2x^2}{l^2} - \frac{1}{2}\right) \quad \left(\text{when } x < \frac{1}{2}\right)$$

$$V = \pm\frac{W}{2}$$

$$M_x = Wx\left(\frac{1}{2} - \frac{x}{l} + \frac{2x^2}{3l^2}\right) \quad \left(\text{when } x < \frac{1}{2}\right)$$

$$M = \frac{Wl}{12} \quad \text{(at center of span)}$$

$$y = \frac{3Wl^3}{320\,EI} \quad \text{(at center of span)}$$

Beam Formulas, continued

Cantilever Beam—Uniform Load

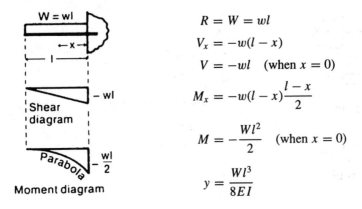

$$R = W = wl$$

$$V_x = -w(l - x)$$

$$V = -wl \quad \text{(when } x = 0\text{)}$$

$$M_x = -w(l - x)\frac{l - x}{2}$$

$$M = -\frac{Wl^2}{2} \quad \text{(when } x = 0\text{)}$$

$$y = \frac{Wl^3}{8EI}$$

Simple Beam—Load Increasing Uniformly from One Support to the Other

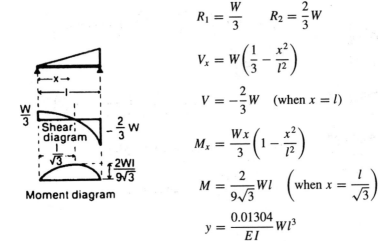

$$R_1 = \frac{W}{3} \qquad R_2 = \frac{2}{3}W$$

$$V_x = W\left(\frac{1}{3} - \frac{x^2}{l^2}\right)$$

$$V = -\frac{2}{3}W \quad \text{(when } x = l\text{)}$$

$$M_x = \frac{Wx}{3}\left(1 - \frac{x^2}{l^2}\right)$$

$$M = \frac{2}{9\sqrt{3}}Wl \quad \left(\text{when } x = \frac{l}{\sqrt{3}}\right)$$

$$y = \frac{0.01304}{EI}Wl^3$$

Beam Formulas, continued

Cantilever Beam—Load Increasing Uniformly from Free End to Support

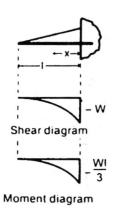

Shear diagram

Moment diagram

$$R = W$$

$$V_x = -W\frac{(l-x)^2}{l^2}$$

$$V = -W \quad (\text{when } x = 0)$$

$$M_x = -\frac{W}{3}\frac{(l-x)^3}{l^2}$$

$$M = -\frac{Wl}{3} \quad (\text{when } x = 0)$$

$$y = \frac{Wl^3}{15EI}$$

Fixed Beam—Concentrated Load at Center of Span

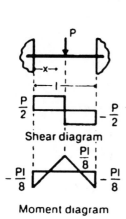

Shear diagram

Moment diagram

$$R_1 = R_2 = \frac{P}{2}$$

$$V_x = V = \pm\frac{P}{2}$$

$$M_x = P\left(\frac{x}{2} - \frac{l}{8}\right) \quad \left(\text{when } x < \frac{l}{2}\right)$$

$$M_x = -\frac{Pl}{8} \quad \left(\text{when } \begin{matrix} x = 0 \\ x = l \end{matrix}\right)$$

$$M = +\frac{Pl}{8} \quad (\text{at center of span})$$

$$y = \frac{Pl^3}{192EI}$$

Beam Formulas, continued

Fixed Beam—Uniform Load

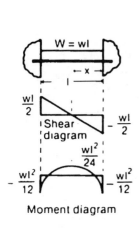

$$R_1 = R_2 = \frac{wl}{2} = \frac{W}{2}$$

$$V_x = \frac{wl}{2} - wx$$

$$V = \pm\frac{wl}{2} \quad \text{(at ends)}$$

$$M_x = -\frac{wl^2}{2}\left(\frac{1}{6} - \frac{x}{l} + \frac{x^2}{l^2}\right)$$

$$M = -\frac{1}{12}wl^2 \quad \left(\text{when } \begin{array}{l} x = 0 \\ x = l \end{array}\right)$$

$$M = \frac{wl^2}{24} \quad \left(\text{when } x = \frac{l}{2}\right)$$

$$y = \frac{Wl^3}{384EI}$$

Simple Beam—Distributed Load over Part of Beam

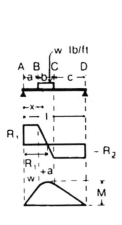

$$R_1 = \frac{wb(2c + b)}{2l}$$

$$R_2 = \frac{wb(2a + b)}{2l}$$

$$V_x = \frac{wb(2c + b)}{2l} - w(x - a)$$

$$V = R_1 \quad \text{(when } a < c)$$

$$\quad = R_2 \quad \text{(when } a > c)$$

$$M_x = \frac{wbx(2c + b)}{2l} \qquad \text{for } AB$$

$$\quad = R_1x - \frac{w(x - a)^2}{2} \qquad \text{for } BC$$

$$\quad = R_2(l - x) \qquad \text{for } CD$$

$$M = \frac{wb(2c + b)[4al + b(2c + b)]}{8l^2}$$

Beam Formulas, continued

Beam Supported at One End, Fixed at Other—Concentrated Load at Any Point

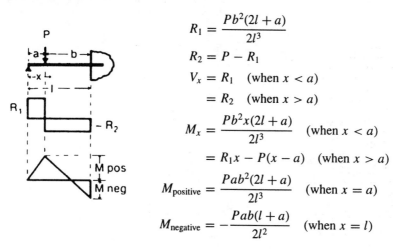

$$R_1 = \frac{Pb^2(2l + a)}{2l^3}$$

$$R_2 = P - R_1$$

$$V_x = R_1 \quad \text{(when } x < a)$$
$$= R_2 \quad \text{(when } x > a)$$

$$M_x = \frac{Pb^2x(2l + a)}{2l^3} \quad \text{(when } x < a)$$

$$= R_1x - P(x - a) \quad \text{(when } x > a)$$

$$M_{\text{positive}} = \frac{Pab^2(2l + a)}{2l^3} \quad \text{(when } x = a)$$

$$M_{\text{negative}} = -\frac{Pab(l + a)}{2l^2} \quad \text{(when } x = l)$$

Fixed Beam—Concentrated Load at Any Point

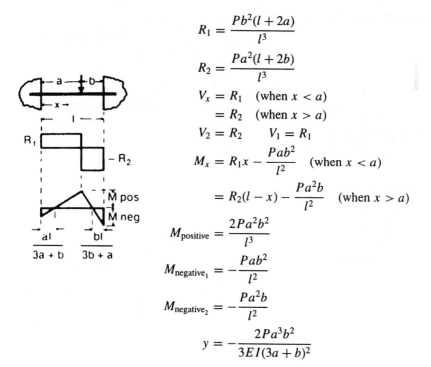

$$R_1 = \frac{Pb^2(l + 2a)}{l^3}$$

$$R_2 = \frac{Pa^2(l + 2b)}{l^3}$$

$$V_x = R_1 \quad \text{(when } x < a)$$
$$= R_2 \quad \text{(when } x > a)$$
$$V_2 = R_2 \qquad V_1 = R_1$$

$$M_x = R_1x - \frac{Pab^2}{l^2} \quad \text{(when } x < a)$$

$$= R_2(l - x) - \frac{Pa^2b}{l^2} \quad \text{(when } x > a)$$

$$M_{\text{positive}} = \frac{2Pa^2b^2}{l^3}$$

$$M_{\text{negative}_1} = -\frac{Pab^2}{l^2}$$

$$M_{\text{negative}_2} = -\frac{Pa^2b}{l^2}$$

$$y = -\frac{2Pa^3b^2}{3EI(3a + b)^2}$$

Beam Formulas, continued

Beam Supported at One End, Fixed at Other—Distributed Load

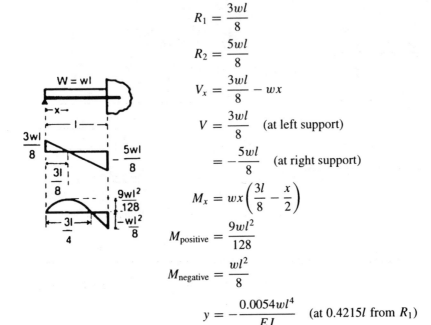

$$R_1 = \frac{3wl}{8}$$

$$R_2 = \frac{5wl}{8}$$

$$V_x = \frac{3wl}{8} - wx$$

$$V = \frac{3wl}{8} \quad \text{(at left support)}$$

$$= -\frac{5wl}{8} \quad \text{(at right support)}$$

$$M_x = wx\left(\frac{3l}{8} - \frac{x}{2}\right)$$

$$M_{\text{positive}} = \frac{9wl^2}{128}$$

$$M_{\text{negative}} = \frac{wl^2}{8}$$

$$y = -\frac{0.0054wl^4}{EI} \quad \text{(at } 0.4215l \text{ from } R_1)$$

Torsion Formulas—Solid and Tubular Sections

Nomenclature

f_s = shear stress, kips/in.^{2} (formulas for f_s apply for stress not exceeding the shear yield strength)

F_{su} = ultimate shear strength, kips/in.^{2}

J = torsion constant,* in.^{4}

T = torque, in.-kips

T_u = approximate ultimate torque, in.-kips

kips = 1000 lb force

*The torsion constant J is a measure of the stiffness of a member in pure twisting.

Torsion Formulas—Solid and Tubular Sections, continued

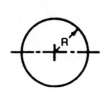

Maximum $f_s = \dfrac{2T}{\pi R^3}$ at boundary

$$T_u = \frac{2\pi R^3 F_{su}}{3}$$

$$J = \frac{\pi R^4}{2}$$

Maximum $f_s = \dfrac{2R_1 T}{\pi \left(R_1^4 - R_2^4\right)}$ at boundary

$$T_u = \frac{2\pi \left(R_1^3 - R_2^3\right) F_{su}}{3}$$

$$J = \frac{\pi}{2}\left(R_1^4 - R_2^4\right)$$

Thin-walled sections should be checked for buckling.

Maximum $f_s = \dfrac{4.8T}{a^3}$ at midpoints of sides

$$T_u = \frac{a^3 F_{su}}{3}$$

$$J = 0.141a^4$$

(continued)

Note: Formulas for C_s are not given in this table because for these cross sections, C_s is negligibly small in comparison to J.

Torsion Formulas—Solid and Tubular Sections, continued

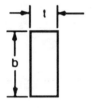

Maximum $f_s = \dfrac{3T}{bt^2}\left(1 + 0.6\dfrac{t}{b}\right)$ at midpoints of long sides

(approximately $3T/bt^2$ for narrow rectangles)

$$T_u = \frac{bt^2 F_{su}}{2}\left(1 - \frac{t}{3b}\right)$$

$$J = \frac{bt^3}{3}\left[1 - 0.63\frac{t}{b} + 0.052\left(\frac{t}{b}\right)^5\right]$$

(approximately $bt^3/3$ for narrow rectangles)

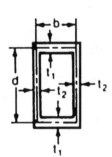

For sides with thickness t_1, average $f_s = \dfrac{T}{2t_1 bd}$

For sides with thickness t_2, average $f_s = \dfrac{T}{2t_2 bd}$

$$T_u = 2bdt_{min}F_{su}$$

($t_{min} = t_1$ or t_2, whichever is smaller)

$$J = \frac{2b^2 d^2}{(b/t_1)(d/t_2)}$$

Thin-walled sections should be checked for buckling.

Note: Formulas for C_s are not given in this table because for these cross sections. C_s is negligibly small in comparison to J.

Torsion Formulas—Thin-Walled Open Sections

Nomenclature

A = total area of section, in.^{2}
A_F = area of one flange, in.^{2}
C_s = torsion-bending constant,* in.^{6}
I_1 = moment of inertia of flange number 1 about Y axis, in.^{4}
I_2 = moment of inertia of flange number 2 about Y axis, in.^{4}
I_Y = moment of inertia of section about Y axis, in.^{4}
J = torsion constant,* in.^{4}

*The torsion constant J is a measure of the stiffness of a member in pure twisting. The torsion-bending constant C_s is a measure of the resistance to rotation that arises because of restraint of warping of the cross section.

Torsion Formulas—Thin-Walled Open Sections, continued

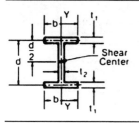

$$J = \frac{1}{3}\left(2bt_1^3 + dt_2^3\right)$$

$$C_s = \frac{d^2 I_Y}{4}$$

$$t_1 \neq t_2$$

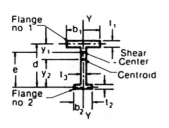

$$J = \frac{1}{3}\left(b_1 t_1^3 + b_2 t_2^3 + dt_3^3\right)$$

$$C_s = \frac{d^2 I_1 I_2}{I_Y}$$

$$e = \frac{y_1 I_1 - y_2 I_2}{I_Y}$$

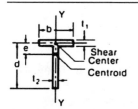

$$J = \frac{1}{3}\left(bt_1^3 + dt_2^3\right)$$

$$C_s = 0$$

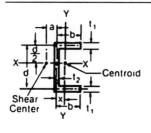

$$J = \frac{1}{3}\left(2bt_1^3 + dt_2^3\right)$$

$$C_s = \frac{d^2 I_Y}{4}\left(1 - \frac{x(a - x)}{r_y^2}\right)$$

$$a = \frac{xd^2}{4r_x^2}$$

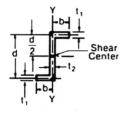

$$J = \frac{1}{3}\left(2bt_1^3 + dt_2^3\right)$$

$$C_s = \frac{d^2 I_Y}{4}\left(1 - \frac{3A_F}{2A}\right)$$

Position of Flexural Center Q for Different Sections

Form of section	Position of Q
Any narrow section symmetrical about the x axis. Centroid at $x = 0$, $y = 0$	$$e = \frac{1 + 3v \int x t^3 dx}{1 + v \int t^3 dx}$$ For narrow triangle (with $v = 0.25$), $$e = 0.187a$$ For any equilateral triangle, $$e = 0$$
Sector of thin circular tube	$$e = \frac{2R}{(\pi - \theta) + \sin\theta\cos\theta}[(\pi - \theta)\cos\theta + \sin\theta]$$ For complete tube split along element ($\theta = 0$), $$e = 2R$$
Semicircular area 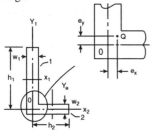	$$e = \left(\frac{8}{15\pi}\frac{3 + 4v}{1 + v}\right)R \quad (Q \text{ is to right of centroid})$$ For sector of solid or hollow circular area
Angle 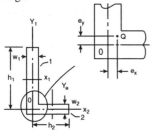	Leg 1 = rectangle $w_1 h_1$; leg 2 = rectangle $w_2 h_2$ I_1 = moment of inertia of leg 1 about Y_1 (central axis) I_2 = moment of inertia of leg 2 about Y_2 (central axis) $$e_x = \frac{1}{2}h_2\left(\frac{I_2}{I_1 + I_2}\right)$$ (For e_x, use x_1 and x_2 central axes.) $$e_y = \frac{1}{2}h_1\left(\frac{I_1}{I_1 + I_2}\right)$$ If w_1 and w_2 are small, $e_x = e_y = 0$ (practically) and Q is at 0.

(continued)

Position of Flexural Center Q for Different Sections, continued

Form of section	Position of Q
Channel	$$e = h\left(\frac{H_{xy}}{I_x}\right)$$ where H_{xy} = product of inertia of the half section (above X) with respect to axes X and Y and I_x = moment of inertia of whole section with respect to X axis If t is uniform, $$e = \frac{b^2 h^2 t}{4 I_x}$$
Tee	$$e = \frac{1}{2}(t_1 + t_2)\left(\frac{1}{1 + \left(d_1^3 t_1 / d_2^3 t_2\right)}\right)$$ For a T-beam of ordinary proportions, Q may be assumed to be at 0.
I with unequal flanges and thin web	$$e = b\left(\frac{I_2}{I_1 + I_2}\right)$$ where I_1 and I_2, respectively, denote moments of inertia about X axis of flange 1 and flange 2
Beam composed of n elements, of any form, connected or separate, with common neutral axis (e.g., multiple-spar airplane wing) 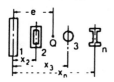	$$e = \frac{E_2 I_2 x_2 + E_3 I_3 x_3 + \cdots + E_n I_n x_n}{E_1 I_1 + E_2 I_2 + E_3 I_3 + \cdots + E_n I_n}$$ where I_1, I_2, etc., are moments of inertia of the several elements about the X axis (i.e., Q is at the centroid of the products $E I$ for the several elements)

(continued)

Position of Flexural Center Q for Different Sections, continued

Form of section		Position of Q				

Lipped channel (t small)

c/h \ b/h	1.0	0.8	0.6	0.4	0.2
0	0.430	0.330	0.236	0.141	0.055
0.1	0.477	0.380	0.280	0.183	0.087
0.2	0.530	0.425	0.325	0.222	0.115
0.3	0.575	0.470	0.365	0.258	0.138
0.4	0.610	0.503	0.394	0.280	0.155
0.5	0.621	0.517	0.405	0.290	0.161

Values of e/h

Hat section (t small)

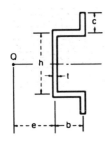

c/h \ b/h	1.0	0.8	0.6	0.4	0.2
0	0.430	0.330	0.236	0.141	0.055
0.1	0.464	0.367	0.270	0.173	0.080
0.2	0.474	0.377	0.280	0.182	0.090
0.3	0.453	0.358	0.265	0.172	0.085
0.4	0.410	0.320	0.235	0.150	0.072
0.5	0.355	0.275	0.196	0.123	0.056
0.6	0.300	0.225	0.155	0.095	0.040

Values of e/h

D section (A = enclosed area)

Values of e(h/A)

t_1/t_s \ S/h	1	1.5	2	3	4	5	6	7
0.5	—	—	1.0	0.800	0.665	0.570	0.500	0.445
0.6	—	—	0.910	0.712	0.588	0.498	0.434	0.386
0.7	—	0.980	0.831	0.641	0.525	0.443	0.384	0.338
0.8	—	0.910	0.770	0.590	0.475	0.400	0.345	0.305
0.9	—	0.850	0.710	0.540	0.430	0.360	0.310	0.275
1.0	1.0	0.800	0.662	0.500	0.400	0.330	0.285	0.250
1.2	0.905	0.715	0.525	0.380	0.304	0.285	0.244	0.215
1.6	0.765	0.588	0.475	0.345	0.270	0.221	0.190	0.165
2.0	0.660	0.497	0.400	0.285	0.220	0.181	0.155	0.135
3.0	0.500	0.364	0.285	0.200	0.155	0.125	0.106	0.091

Buckling and Stress

Shear-Buckling Curves for Sheet Materials

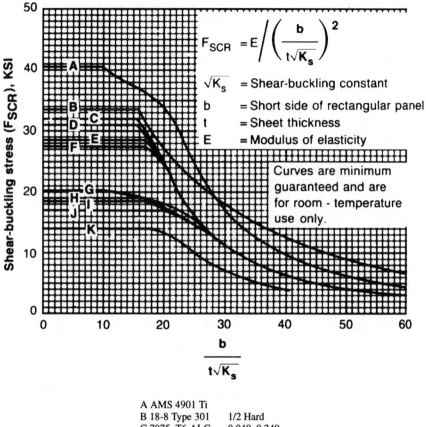

$$F_{SCR} = E \Big/ \left(\frac{b}{t\sqrt{K_s}} \right)^2$$

$\sqrt{K_s}$ = Shear-buckling constant

b = Short side of rectangular panel

t = Sheet thickness

E = Modulus of elasticity

Curves are minimum guaranteed and are for room - temperature use only.

A AMS 4901 Ti
B 18-8 Type 301   1/2 Hard
C 7075–T6 ALC.   0.040–0.249
D 7075–T6 ALC.   0.012–0.039
E 2024–T81 ALC.   t≥0.063
F 2024–T81 ALC.   t≤0.063
G 6061–T6   t≥0.250
H 2024–T3 ALC.   0.063–0.249
I 2024–T3 ALC.   0.010–0.062
J 2024–T42 ALC.   t<0.063
K AZ31B–H24   0.016–0.250

Source: Vought Aerospace Corporation handbook. Copyright © Northrop Grumman, Los Angeles. Reproduced with permission of Northrop Grumman.

Buckling and Stress, continued

Allowable Shear Flow—7075-T6 Clad Web

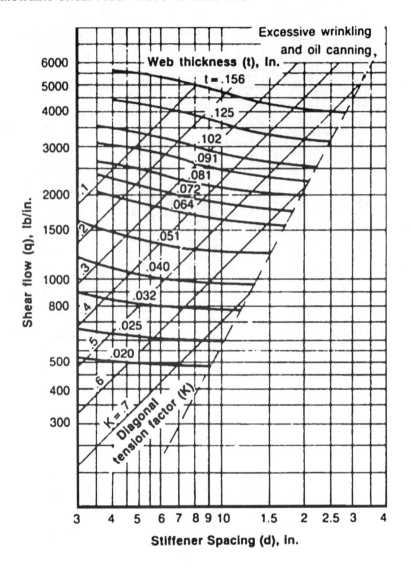

Buckling and Stress, continued

Shear-Buckling Constant ($\sqrt{K_s}$)

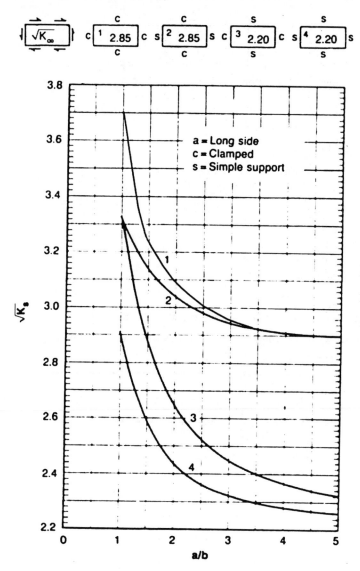

Source: Vought Aerospace Corporation handbook. Copyright © Northrop Grumman, Los Angeles. Reproduced with permission of Northrop Grumman.

Buckling and Stress, continued

Beam Diagonal Tension Nomenclature

A = cross-sectional area, in.^{2}
d = spacing of uprights, in.
q = shear flow (shear force per in.), kips per in.
t = thickness, in. (when used without subscript, signifies thickness of web)
k = diagonal-tension factor
h = depth of beam, in.
h_U = length of upright (measured between centroids of upright-to-flange
       rivet patterns), in.
σ = normal stress, ksi
τ = shear stress, ksi
ωd = flange flexibility factor
α = angle between neutral axis of beams and direction of diagonal tension, deg

Subscripts:

DT = diagonal tension
IDT = incomplete diagonal tension
PDT = pure diagonal tension
U = upright
e = effective

Maximum Stress to Average Stress in Web Stiffener

For curved webs: for rings, read abscissa as d/h; for stringers, read abscissa as h/d.

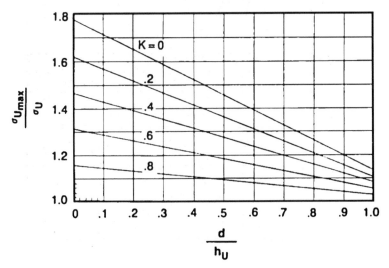

Source: The diagonal tension material appearing on pages 4-20 and 4-21 is from *Summary of Diagonal Tension, Part I*, by NACA, NACA-TN-2661.

Buckling and Stress, continued

Angle Factor C_1

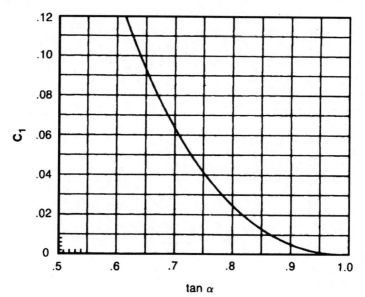

tan α

Stress-Concentration Factors C_2 and C_3

$$\left(\omega d = 0.7d \sqrt[4]{\frac{t}{(I_c + I_T)h_c}} \right)$$

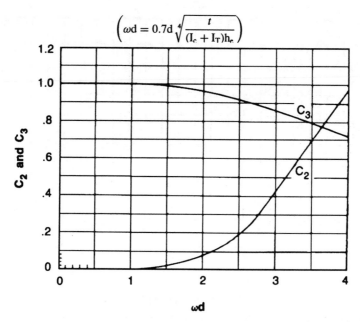

ωd

$I_C = I$ of compression flange.

$I_T = I$ of tension flange.

Buckling and Stress, continued

Diagonal-Tension Analysis Chart

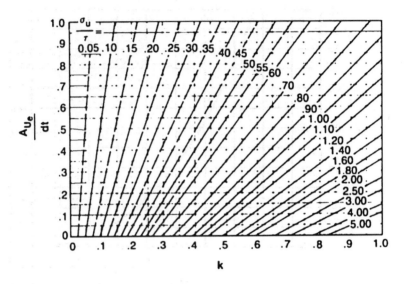

Angle of Diagonal Tension

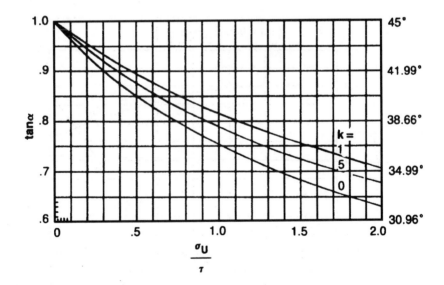

Incomplete diagonal tension

Buckling and Stress, continued
Angle of Diagonal Tension, continued

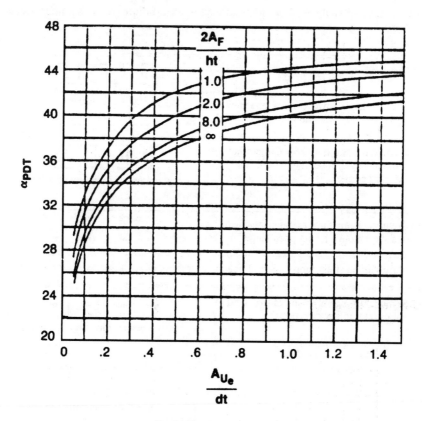

Pure diagonal tension

Columns

Interaction of Column Failure with Local Failure (Crippling)

The method of analysis of columns subject to local failure can be summarized as follows:

a) Sections having four corners	▯ ⌐ ⌐ ⊔
b) Sections having two corners, attached to sheets along both flanges	I ⊏
c) Sections having only two corners but restrained against column failure about axis thru corners	⌐↑ ⊏↑ ⌐↑ I↑ (Arrow represents direction of column failure)
d) Sections having three corners, attached to a sheet along one unlipped flange	⌐ ⌐

For local failing stress (upper limit of column curve) use crippling stress F_{cc} (see note)

e) Sections having only two corners (with no restraint in any direction)	⌐→ ⊏→ ⌐→I→ (Arrow represents direction of column failure)
f) Sections having only two corners, attached to sheet	⌐→ ⊏→ ⊓↓ (Arrow represents direction of column failure)

For local failing stress (upper limit of column curve) use local buckling stress F_{ccr} (see note)

Column curve	For columns that fail by Euler buckling, select the proper column curve and determine the allowable ultimate stress F_C.

Note: The calculations for F_{cc} and F_{ccr} should be made by reference to sources not included in this handbook.

Source: Northrop Grumman structures manual. Copyright © Northrop Grumman Corporation, Los Angeles. Reproduced with permission of Northrop Grumman.

Columns, continued

Comparison of Different Column Curves

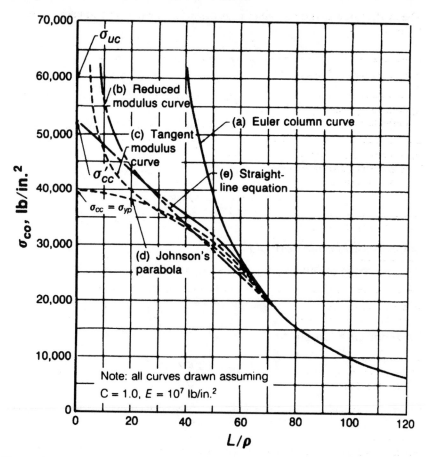

Euler Column Formula

$$P_E = \frac{C\pi^2 EI}{L^2}$$

where P_E = critical column load and C = end fixity coefficient (pin end = 1.0, restrained = 4.0). The critical column stress σ_E is

$$\sigma_E = \frac{C\pi^2 E}{(L/\rho)^2}$$

where ρ = radius of gyration = $\sqrt{I/A}$.

The graph and text appearing on pages 4-25 and 4-26 are from *Weight Engineers Handbook*, Revised 1976. Copyright © 1976, the Society of Allied Weight Engineers. Used with permission of the Society of Allied Weight Engineers.

Columns, continued

Reduced Modulus Curve

The L/ρ corresponding to the critical stress σ_{sc} in the short column range is

$$\left(\frac{L}{\rho}\right)_{E_r} = \pi\sqrt{\frac{E_r}{\sigma_{sc}}}$$

where

$$E_r = \left(\sqrt{\frac{4EE_t}{\sqrt{E}+\sqrt{E_t}}}\right)^2 \qquad E_r = \sqrt{EE_t} \qquad E_t = \text{tangent modulus}$$

Tangent Modulus Curve

$$\left(\frac{L}{\rho}\right)_{E_t} = \pi\sqrt{\frac{E_t}{\sigma_{sc}}}$$

Johnson Parabolic Formula

The Johnson equation gives the critical short column stress σ_{sc} is

$$\sigma_{sc} = \sigma_{cc} - \frac{\sigma_{cc}^2(L/\rho)^2}{4C\pi^2 E}$$

Straight Line Equation

$$\sigma_{sc} = \sigma_{cc}'\frac{1-k(L/\rho)}{\sqrt{C}}$$

where σ_{cc}, k, and $\sqrt{C}$ are chosen to give best agreement with experimental data.

Columns, continued

Column Curves for Aluminum Alloys—Based on Tangent Modulus

For thick sections-no local crippling

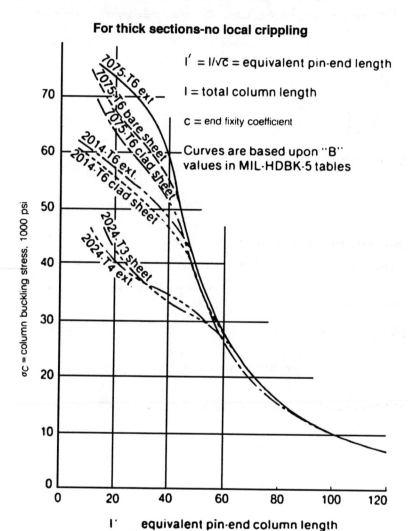

$l' = l/\sqrt{c}$ = equivalent pin-end length

l = total column length

c = end fixity coefficient

Curves are based upon "B" values in MIL-HDBK-5 tables

Curves labeled: 7075-T6 ext, 7075-T6 bare sheet, 7075-T6 clad sheet, 2014-T6 ext, 2014-T6 clad sheet, 2024-T3 sheet, 2024-T4 ext

σ_c = column buckling stress, 1000 psi

$$\frac{l'}{\rho} = \frac{\text{equivalent pin-end column length}}{\text{section radius of gyration}}$$

Source: Northrop Grumman structures manual. Copyright © Northrop Grumman Corporation, Los Angeles. Reproduced with permission of Northrop Grumman.

Columns, continued

End Fixity Coefficients

Column shape and end conditions	End fixity coefficient
Uniform column, axially loaded, pinned ends	$c = 1$ $\dfrac{1}{\sqrt{c}} = 1$
Uniform column, axially loaded, fixed ends	$c = 4$ $\dfrac{1}{\sqrt{c}} = 0.5$
Uniform column, axially loaded, one end fixed, one end pinned	$c = 2.05$ $\dfrac{1}{\sqrt{c}} = 0.70$
Uniform column, axially loaded, one end fixed, one end free	$c = 0.25$ $\dfrac{1}{\sqrt{c}} = 2$

(continued)

Columns, continued

End Fixity Coefficients, continued

Column shape and end conditions	End fixity coefficient
Uniform column, distributed axial load, one end fixed, one end free	$c = 0.794$ $\dfrac{1}{\sqrt{c}} = 1.12$
Uniform column, distributed axial load, pinned ends	$c = 1.87$ $\dfrac{1}{\sqrt{c}} = 0.732$
Uniform column, distributed axial load, fixed ends	$c = 7.5$ $\dfrac{1}{\sqrt{c}} = 0.365$
Uniform column, distributed axial load, one end fixed, one end pinned	$c = 3.55$ (approx.) $\dfrac{1}{\sqrt{c}} = 0.530$

Columns, continued

Column Stress for Aluminum Alloy Columns

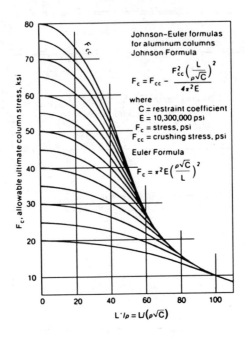

Johnson-Euler formulas for aluminum columns

Johnson Formula

$$F_c = F_{cc} - \frac{F_{cc}^2 \left(\frac{L}{\rho\sqrt{C}}\right)^2}{4\pi^2 E}$$

where
  C = restraint coefficient
  E = 10,300,000 psi
  F_c = stress, psi
  F_{cc} = crushing stress, psi

Euler Formula

$$F_c = \pi^2 E \left(\frac{\rho\sqrt{C}}{L}\right)^2$$

$L' / \rho = L / (\rho\sqrt{C})$

(y-axis: F_c, allowable ultimate column stress, ksi)

Column Stress for Magnesium Alloy Columns

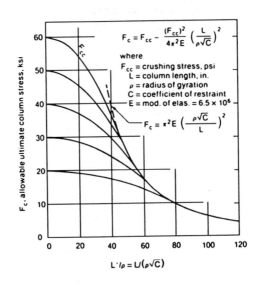

$$F_c = F_{cc} - \frac{(F_{cc})^2}{4\pi^2 E} \left(\frac{L}{\rho\sqrt{C}}\right)^2$$

where
  F_{cc} = crushing stress, psi
  L = column length, in.
  ρ = radius of gyration
  C = coefficient of restraint
  E = mod. of elas. = 6.5×10^6

$$F_c = \pi^2 E \left(\frac{\rho\sqrt{C}}{L}\right)^2$$

$L' / \rho = L / (\rho\sqrt{C})$

(y-axis: F_c, allowable ultimate column stress, ksi)

Columns, continued

Column Stress for Steel Columns

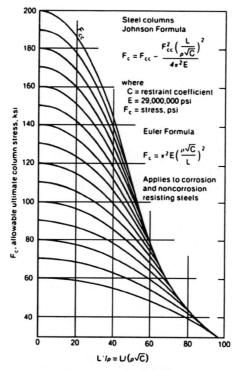

Steel columns
Johnson Formula

$$F_c = F_{cc} - \frac{F_{cc}^2 \left(\frac{L}{\rho\sqrt{C}}\right)^2}{4\pi^2 E}$$

where
 C = restraint coefficient
 E = 29,000,000 psi
 F_c = stress, psi

Euler Formula

$$F_c = \pi^2 E \left(\frac{\rho\sqrt{C}}{L}\right)^2$$

Applies to corrosion
and noncorrosion
resisting steels

$L'/\rho = L/(\rho\sqrt{C})$

Column Stress for Titanium Alloy Columns

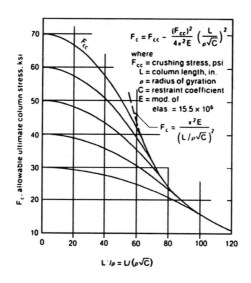

$$F_c = F_{cc} - \frac{(F_{cc})^2}{4\pi^2 E} \left(\frac{L}{\rho\sqrt{C}}\right)^2$$

where
 F_{cc} = crushing stress, psi
 L = column length, in.
 ρ = radius of gyration
 C = restraint coefficient
 E = mod. of
   elas. = 15.5×10^6

$$F_c = \frac{\pi^2 E}{\left(L/\rho\sqrt{C}\right)^2}$$

$L'/\rho = L/(\rho\sqrt{C})$

Columns, continued

Beam-Column Curves for Centrally Loaded Columns with End Couples

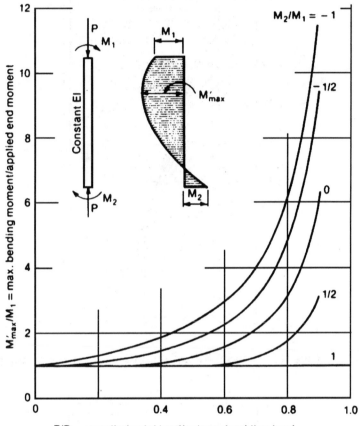

P/P$_{cr}$ = applied axial load/column buckling load

P   = applied axial load
P$_{cr}$ = F$_c$ × A = column buckling load
F$_c$ = column buckling stress
M$_1$, M$_2$ = applied end moment
M$'_{max}$ = max. total bending moment including secondaries

Source: Westergaard, H. M., "Buckling of Elastic Structure," Trans. ASCE, 1922, p. 594.

Plates

Buckling Stress Coefficients for Flat Plates in Shear

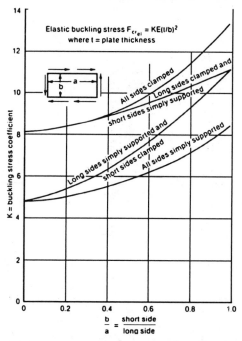

Sources: TN 1222, 1559; Timoshenko, "Theory of Elastic Stability;" RAS Data Sheets; "Stressed Skin Structures."

Buckling Stress Coefficients for Flat Rectangular Plates Loaded in Compression on the Long Side

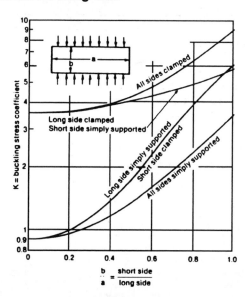

Plates, continued

Buckling Stress Coefficients for Flat Rectangular Plates Loaded in Compression on the Short Side

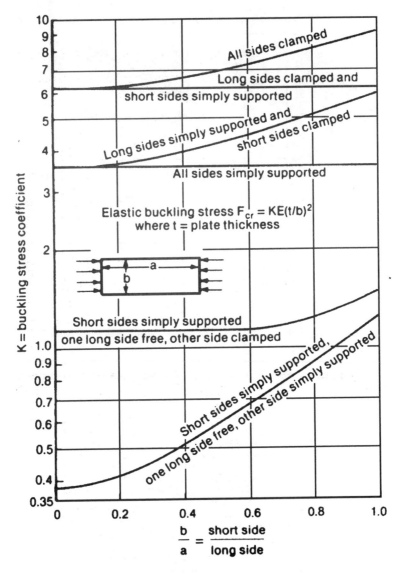

Sources: Timoshenko, "Theory of Elastic Stability;" RAS Data Sheets; "Stressed Skin Structures."

Plates, continued

Stress Concentration Factors for an Axially Loaded Plate with an Eccentric Circular Hole

Note: These factors apply to stresses in the elastic region

$$f_{max} = K \times f_{net} \quad = \text{maximum stress (at point "n")}$$

$$f_{net} = P/(d_1 + d_2)(t) = \text{average net section stress}$$
$$t = \text{plate thickness}$$

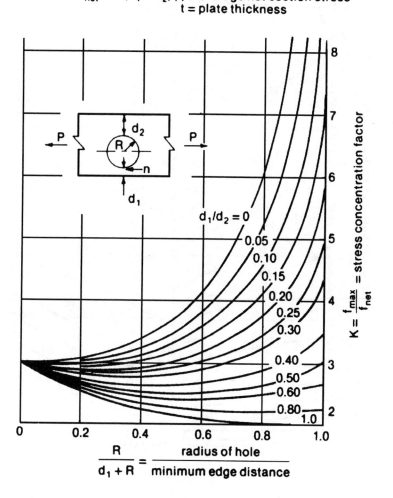

Source: Aero. Res. Inst. of Sweden, Rept. No. 36.

Plates, continued

Stress Concentration Factors for a Flat Plate with a Shoulder Fillet under Axial Load

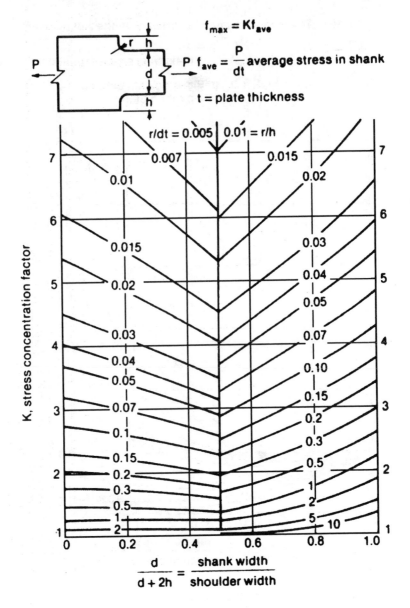

$$f_{max} = Kf_{ave}$$

$$P \quad f_{ave} = \frac{P}{dt} \text{ average stress in shank}$$

$t = $ plate thickness

$$\frac{d}{d + 2h} = \frac{\text{shank width}}{\text{shoulder width}}$$

Sources: Peterson, R. E., "Stress Concentration Factors," Fig. 57; TN 2442; GAEC Data.

Rivet Strengths

Static joint strength of 100° flush head aluminum alloy solid rivets in machine-countersunk aluminum alloy sheet

Sheet material	Clad 2024-T42								
Rivet type	MS20426AD (2117-T3) (F_{su} = 30 ksi)				MS20426D (2017-T3) (F_{su} = 38 ksi)			MS20426DD (2024-T31) (F_{su} = 41 ksi)	
Rivet diameter, in. (Nominal hole diameter, in.)	3/32 (0.096)	1/8 (0.1285)	5/32 (0.159)	3/16 (0.191)	5/32 (0.159)	3/16 (0.191)	1/4 (0.257)	3/16 (0.191)	1/4 (0.257)
Sheet thickness, in.	Ultimate Strength,[a] lb								
0.032	178	—	—	—					
0.040	193	309[c]	—	—					
0.050	206	340	479[c]	—	580[b]	—	—		
0.063	216	363	523	705[c]	657[b]	859[b,c]	—		
0.071	—	373	542	739	690[c]	917[b]	—	886	
0.080	—	—	560	769	720	969[b]	—	942	
0.090	—	—	575	795	746	1015	—	992	1647[b]
0.100	—	—	—	818	—	1054	1552[b]	1035	1738[b,c]
0.125	—	—	—	853	—	1090	1640[b]	1073[c]	1877
0.160	—	—	—	—	—	—	1773	1131	2000
0.190	—	—	—	—	—	—	1891	—	2084
Rivet shear strength[d]	217	388	596	862	755	1090	1970	1175	2125

[a] Test data from which the yield and ultimate strength listed were derived can be found in "Report on Flush Riveted Joint Strength" (Airworthiness Requirements Committee, A/C Industries Association of America, Inc., Airworthiness Project 12 [Revised May 25, 1948]).

[b] Yield value is less than 2/3 of the indicated ultimate strength value.

[c] Values above line are for knife-edge condition, and the use of fasteners in this condition is undesirable. The use of knife-edge condition in design of military aircraft requires specific approval of the procuring activity.

[d] Rivet shear strength is documented in MS20426.

Source: MIL-HDBK-5G.

Bolt and Screw Strengths

Tension and shear values

| | Minimum ultimate load, lb | | | |
| | 160 ksi tensile | | 180 ksi tensile | |
Diameter and thread UNJF-3A	95 ksi double shear[a]	Flush tension and pan head[b]	108 ksi double shear[a]	Flush tension and pan head[b]
0.1900-32	5,400	3,180	6,120	3,600
0.2500-28	9,330	5,280	10,600	6,500
0.3125-24	14,600	9,200	16,500	10,400
0.3750-24	21,000	14,000	23,800	15,800
0.4375-20	28,600	18,900	32,400	21,300
0.5000-20	37,300	25,600	42,400	28,700
0.5625-18	47,200	32,400	53,600	36,500
0.6250-18	58,300	41,000	66,200	46,000
0.7500-16	83,900	59,500	95,400	63,200
0.8750-14	114,000	81,500	130,000	86,300
1.0000-12	149,000	106,000	170,000	112,000
1.1250-12	189,000	137,000	214,000	144,000
1.2500-12	233,000	171,000	266,000	180,000

| | Minimum ultimate load, lb | | | | | |
| | 220 ksi tensile | | | 260 ksi tensile | | |
Diameter and thread UNJF-3A	125 ksi double shear[a]	132 ksi double shear[a]	Flush tension and pan head[c]	145 ksi double shear[a]	156 ksi double shear[a]	Flush tension and pan head[d]
0.1900-32	7,080	7,480	3,910	8,220	8,840	4,560
0.2500-28	12,300	13,900	6,980	14,200	15,300	8,150
0.3125-24	19,200	20,200	11,100	22,200	24,000	12,900
0.3750-24	27,600	29,200	17,100	32,000	34,400	20,000
0.4375-20	37,600	39,600	23,200	43,600	47,000	27,000
0.5000-20	49,000	51,800	30,900	57,000	61,200	36,100
0.5625-18	62,200	65,600	39,200	72,000	77,600	45,700
0.6250-18	76,600	81,000	49,000	89,000	95,800	57,200
0.7500-16	110,000	117,000	71,100	128,000	138,000	83,000
0.8750-14	150,000	159,000	97,100	174,000	188,000	113,000
1.0000-12	196,000	208,000	126,000	228,000	246,000	148,000
1.1250-12	248,000	262,000	162,000	288,000	310,000	189,000
1.2500-12	306,000	324,000	202,000	356,000	382,000	236,000

[a]Check that hole material can develop full bearing strength. Ref.: MIL-B-87114A.
[b]Based on FED-STD-H28 tensile stress areas.
[c]Based on 180 ksi multiplied by NAS1348 tensile stress areas.
[d]Based on 210 ksi multiplied by NAS1348 tensile stress areas.

Bolt and Screw Strengths, continued

Summary of fastener materials

Material	Surface treatment	Useful design temperature limit, °F	Ultimate tensile strength at room temperature, ksi	Comments
Carbon steel	Zinc plate	−65 to 250	55 and up	——
Alloy steels	Cadmium plate, nickel plate, zinc plate, or chromium plate	−65 to limiting temperature of plating	Up to 300	Some can be used at 900°F
A-286 stainless	Passivated per MIL-S-5002	−423 to 1200	Up to 220	——
17-4PH stainless	None	−300 to 600	Up to 220	——
17-7PH stainless	Passivated	−200 to 600	Up to 220	——
300 series stainless	Furnace oxidized	−423 to 800	70 to 140	Oxidation reduces galling
410, 416, and 430 stainless	Passivated	−250 to 1200	Up to 180	47 ksi at 1200°F; will corrode slightly
U-212 stainless	Cleaned and passivated per MIL-S-5002	1200	185	140 ksi at 1200°F
Inconel 718 stainless	Passivated per QQ-P-35 or cadmium plated	−423 to 900 or cadmium plate limit	Up to 220	——
Inconel X-750 stainless	None	−320 to 1200	Up to 180	136 ksi at 1200°F
Waspalloy stainless	None	−423 to 1600	150	——
Titanium	None	−350 to 500	Up to 160	——

Source: Barrett, R. T., "Fastener Design Manual," NASA Reference Publication 1228, Lewis Research Center, Cleveland, OH, 1990.

Vibration

Natural Frequencies

Mass-Spring Systems in Translation (Rigid Mass and Massless Spring)

k = spring stiffness, lb/in.

m = mass, lb-s^2/in.

ω_n = angular natural frequency, rad/s

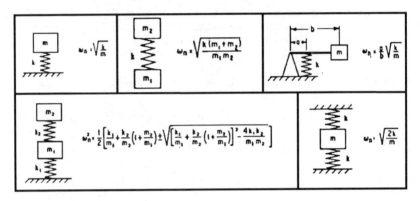

Springs in Combination

k_r = resultant stiffness of combination

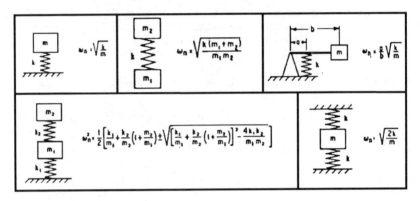

Characteristics of Harmonic Motion

A body that experiences simple harmonic motion follows a displacement pattern defined by

$$x = x_0 \sin(2\pi f t) = x_0 \sin \omega t$$

where f is the frequency of the simple harmonic motion, $\omega = 2\pi f$ is the corresponding angular frequency, and x_0 is the amplitude of the displacement.

Characteristics of Harmonic Motion, continued

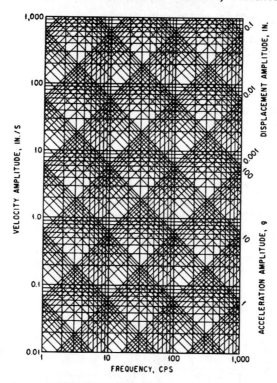

Relation of frequency to the amplitudes of displacement, velocity, and acceleration in harmonic motion. The acceleration amplitude is expressed as a dimensionless multiple of the gravitational acceleration g, where $g = 386$ in./s^2.

Forced Vibration

Single Degree of Freedom

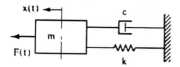

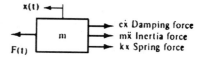

The equation of motion of the system is

$$m\ddot{x} + c\dot{x} + kx = F_0 \sin \omega t$$

where

m = mass
c = viscous damping constant
k = spring stiffness
$F(t) = F_0 \sin \omega t$ = sinusoidal external force
ω = forcing frequency

Forced Vibration, continued

Single Degree of Freedom, continued

The steady-state solution is

$$x = a \sin(\omega t - \theta)$$

where

$$a = \frac{a_s}{\sqrt{\left(1 - \omega^2/\omega_n^2\right)^2 + [2\xi(\omega/\omega_n)]^2}}$$

$$\tan \theta = \frac{2\xi(\omega/\omega_n)}{1 - (\omega/\omega_n)^2}$$

a_s = F_0/k = maximum static-load displacement
θ = phase angle
ξ = c/c_c = fraction of critical damping
c_c = $2\sqrt{mk}$ = $2m\omega_n$ = critical damping
ω_n = $\sqrt{k/m}$, natural frequency, rad/s

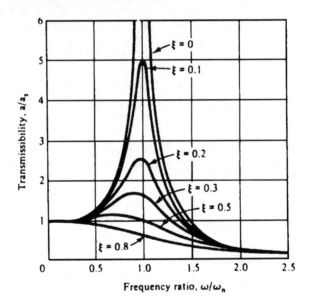

Forced response of a single-degree-of-freedom system

Forced Vibration, continued

Single Degree of Freedom, continued

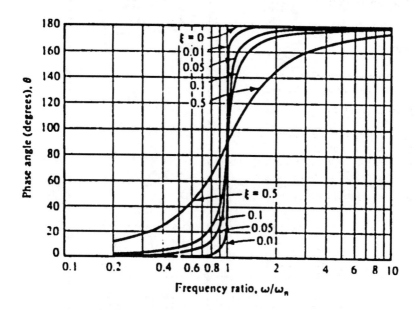

Forced response of a single-degree-of-freedom system

Beam Natural Frequency

Massive Springs (Beams) with Concentrated Mass Loads

m = mass of load, lb-s^2/in.
$m_s(m_b)$ = mass of spring (beam), lb-s^2/in.
k = stiffness of spring, lb/in.
l = length of beam, in.
I = area moment of inertia of beam cross section, in.
E = Young's modulus, lb/in.^{2}
ω_n = angular natural frequency, rad/s

Beam Natural Frequency, continued

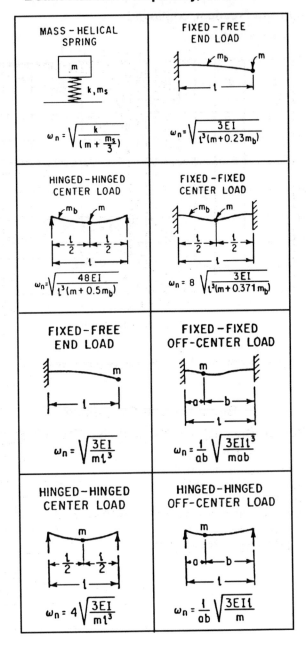

MASS – HELICAL SPRING

$$\omega_n = \sqrt{\frac{k}{(m + \frac{m_s}{3})}}$$

FIXED – FREE END LOAD

$$\omega_n = \sqrt{\frac{3EI}{l^3(m + 0.23m_b)}}$$

HINGED – HINGED CENTER LOAD

$$\omega_n = \sqrt{\frac{48EI}{l^3(m + 0.5m_b)}}$$

FIXED – FIXED CENTER LOAD

$$\omega_n = 8\sqrt{\frac{3EI}{l^3(m + 0.371m_b)}}$$

FIXED – FREE END LOAD

$$\omega_n = \sqrt{\frac{3EI}{ml^3}}$$

FIXED – FIXED OFF-CENTER LOAD

$$\omega_n = \frac{1}{ab}\sqrt{\frac{3EIl^3}{mab}}$$

HINGED – HINGED CENTER LOAD

$$\omega_n = 4\sqrt{\frac{3EI}{ml^3}}$$

HINGED – HINGED OFF-CENTER LOAD

$$\omega_n = \frac{1}{ab}\sqrt{\frac{3EIl}{m}}$$

Beam Natural Frequency, continued

ω_n = angular natural frequency, $A\sqrt{\frac{EI}{\mu l^4}}$ rad/s

E = Young's modulus, lb/in.^{2}

I = area moment of inertia of beam cross section, in.^{4}

l = length of beam, in.

μ = mass per unit length of beam, lb-s^2/in.^{2}

A = coefficient from the following table

Nodes are indicated in the following table as a proportion of length l measured from left end.

	First mode	Second mode
Fixed–free (cantilever)	$A = 3.52$	0.774 $A = 22.4$
Hinged–hinged (simple)	$A = 9.87$	0.500 $A = 39.5$
Fixed–fixed (built-in)	$A = 22.4$	0.500 $A = 61.7$
Free–free	0.224 0.776 $A = 22.4$	0.132 0.500 0.868 $A = 61.7$
Fixed–hinged	$A = 15.4$	0.560 $A = 50.0$
Hinged–free	0.736 $A = 15.4$	0.446 0.853 $A = 50.0$

Plate Natural Frequency

$\omega_n = 2\pi f_n$
$D = Eh^3/12(1 - \mu^2)$
γ = weight density
h = plate thickness
s = denotes simply supported edge
c = denotes built-in or clamped edge
a = length of plate
b = width of plate

s-s-s-s	b/a	1.0	1.5	2.0	2.5	3.0	∞
	$\omega_n/\sqrt{Dg/\gamma ha^4}$	19.74	14.26	12.34	11.45	10.97	9.87

c-s-s-s	b/a	1.0	1.5	2.0	2.5	3.0	∞
	$\omega_n/\sqrt{Dg/\gamma ha^4}$	23.65	18.90	17.33	16.63	16.26	15.43
	a/b	1.0	1.5	2.0	2.5	3.0	∞
	$\omega_n/\sqrt{Dg/\gamma ha^4}$	23.65	15.57	12.92	11.75	11.14	9.87

c-s-c-s	b/a	1.0	1.5	2.0	2.5	3.0	∞
	$\omega_n/\sqrt{Dg/\gamma ha^4}$	28.95	25.05	23.82	23.27	22.99	22.37
	a/b	1.0	1.5	2.0	2.5	3.0	∞
	$\omega_n/\sqrt{Dg/\gamma ha^4}$	28.95	17.37	13.69	12.13	11.36	9.87

c-c-c-c	b/a	1.0	1.5	2.0	2.5	3.0	∞
	$\omega_n/\sqrt{Dg/\gamma ha^4}$	35.98	27.00	24.57	23.77	23.19	22.37

Section 5

MECHANICAL DESIGN

Springs

Spring Nomenclature

A = cross-section area of wire, in.^{2}
C = spring index = D/d
CL = compressed length, in.
D = mean coil diameter, in.
d = diameter of wire or side of square, in.
E = elastic modulus (Young's modulus), psi
FL = free length, unloaded spring, in.
f = stress, tensile or compressive, psi
G = shear modulus of elasticity in torsion, psi = $E/(2[1 + \mu])$
IT = initial tension
in. = inch
J = torsional constant, in.^{4}
k = spring rate, lb/in. = P/s
L = active length subject to deflection, in.
l = length, in.
lb = pound
N = number of active coils
OD = outside diameter, in.
P = load, lb
P_1 = applied load, lb (also P_2, etc.)
p = pitch, in.
psi = pounds per square inch
r = radius of wire, in.
SH = solid height, in.
s = deflection, in.
T = torque, in.-lb
TC = total number of coils
t = leaf spring thickness, in.
U = elastic energy (strain energy), in.-lb
w = leaf spring width, in.
μ = Poisson's ratio
π = pi, 3.1416
τ = stress, torsional, psi
τ_{it} = stress, torsional, due to initial tension, psi

Springs, continued

Maximum shear stress in wire, τ:

$$\tau = \frac{Tr}{J} + \frac{P}{A}$$

where

$$A = \frac{\pi d^2}{4} \qquad T = \frac{PD}{2} \qquad r = \frac{d}{2} \qquad J = \frac{\pi d^4}{32}$$

$$\tau = \frac{(PD/2)(d/2)}{\pi d^4/32} + \frac{P}{\pi d^2/4}$$

Let $C = D/d$ and $K_s = 1 + 1/2C$, then

$$\tau = \frac{8PD}{\pi d^3}\left(1 + \frac{1}{2C}\right) = K_s\frac{8PD}{\pi d^3}$$

Spring deflection:

$$U = \frac{T^2 l}{2GJ} = \frac{4P^2 D^3 N}{d^4 G}$$

$$s = \frac{\partial U}{\partial P} = \frac{8PD^3 N}{d^4 G} \qquad \text{(Castigliano's Theorem)}$$

Helical Springs

Lateral buckling of compression springs

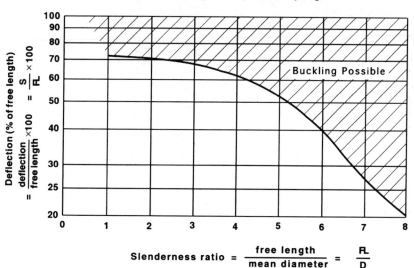

Deflection (% of free length)

$\dfrac{S}{FL} \times 100$

$= \dfrac{\text{deflection} \times 100}{\text{free length}}$

Slenderness ratio $= \dfrac{\text{free length}}{\text{mean diameter}} = \dfrac{FL}{D}$

MECHANICAL DESIGN

Springs, continued

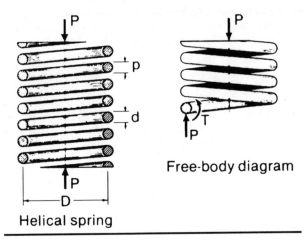

Helical spring

Free-body diagram

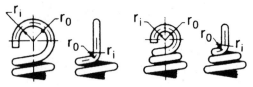

Extension spring ends

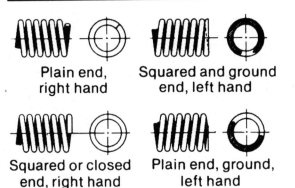

Plain end,
right hand

Squared and ground
end, left hand

Squared or closed
end, right hand

Plain end, ground,
left hand

Compression spring ends

Springs, continued

Preferred sizes for spring materials—wire, strip, and bars

	Spring steels		Corrosion resisting		Copper and nickel alloys	
Music wire	High carbon and alloy steels	Valve spring quality steels	18-8 chrome nickel austenitic 300 series	Straight chrome martensitic 400 series	Spring quality brass phosphor bronze beryllium cop. monel and inconel	K monel and inconel X-750
0.004	0.032	0.092	0.004	Same as high	0.010	0.125
0.006	0.035	0.105	0.006	carbon and	0.012	0.156
0.008	0.041	0.125	0.008	alloy steels,	0.014	0.162
0.010	0.047	0.135	0.010	Col. 2	0.016	0.188
0.012	0.054	0.148	0.012		0.018	0.250
0.014	0.063	0.156	0.014		0.020	0.313
0.016	0.072	0.162	0.020		0.025	0.375
0.018	0.080	0.177	0.026		0.032	0.475
0.020	0.092	0.188	0.032		0.036	0.500
0.022	0.105	0.192	0.042		0.040	0.563
0.024	0.125	0.207	0.048		0.045	0.688
0.026	0.135	0.218	0.054		0.051	0.750
0.028	0.148	0.225	0.063		0.057	0.875
0.032	0.156	0.244	0.072		0.064	1.000
0.042	0.162	0.250	0.080		0.072	1.125
0.048	0.177		0.092		0.081	1.250
0.063	0.188		0.105		0.091	1.375
0.072	0.192		0.120		0.102	1.500
0.080	0.207		0.125		0.114	1.625
0.090	0.218		0.135		0.125	1.750
0.107	0.225		0.148		0.128	2.000
0.130	0.244		0.156		0.144	
0.162	0.250		0.162		0.156	
0.177	0.263		0.177		0.162	
	0.283		0.188		0.182	
	0.307		0.192		0.188	
	0.313		0.207		0.250	
	0.362		0.218			
	0.375		0.225			
			0.250			
			0.312			
			0.375			

Springs, continued

Formulas for compression and extension springs

Property	Round wire	Square wire
Torsional stress (τ), psi	$\dfrac{PD}{0.393d^3}$	$\dfrac{PD}{0.416d^3}$
	$\dfrac{Gds}{\pi ND^2}$	$\dfrac{Gds}{2.32ND^2}$
Deflection (s), in.	$\dfrac{8PND^3}{Gd^4}$	$\dfrac{5.58PND^3}{Gd^4}$
	$\dfrac{\pi \tau ND^2}{Gd}$	$\dfrac{2.32\tau ND^2}{Gd}$
Change in load $(P_2 - P_1)$, lb	$(L_1 - L_2)k$	$(L_1 - L_2)k$
Change in load $(P_1 - P_2)$, lb	$(L_2 - L_1)k$	$(L_2 - L_1)k$
Stress due to initial tension (τ_{it}), psi	$\dfrac{\tau}{P}IT$	$\dfrac{\tau}{P}IT$
Rate (k), lb/in.	P/s	P/s

Compression spring dimensional characteristics

Dimensional characteristics	Type of ends			
	Open or plain (not ground)	Open or plain with ends ground	Squared or closed (not ground)	Closed and ground
Pitch (p)	$\dfrac{FL - d}{N}$	$\dfrac{FL}{TC}$	$\dfrac{FL - 8d}{N}$	$\dfrac{FL - 2d}{N}$
Solid height (SH)	$(TC + 1)d$	$TC \times d$	$(TC + 1)d$	$TC \times d$
Active coils (N)	$N - TC$	$N - TC - 1$	$N - TC - 2$	$N - TC - 2$
	or	or	or	or
	$\dfrac{FL - d}{p}$	$\dfrac{FL}{p} - 1$	$\dfrac{FL - 8d}{p}$	$\dfrac{FL - 2d}{p}$
Total coils (TC)	$\dfrac{FL - d}{p}$	$\dfrac{FL}{p}$	$\dfrac{FL - 8d}{p} + 2$	$\dfrac{FL - 2d}{p} + 2$
Free length (FL)	$(p \times TC) + d$	$p \times TC$	$(p \times N) + 3d$	$(p \times N) + 2d$

Springs, continued

Spring Rate *K* and Spring Work *U* (Strain Energy)

Characteristic		
	General	Const.
k	$\dfrac{\mathrm{d}P}{\mathrm{d}s}$	$\dfrac{P}{s}$
U_1	$\displaystyle\int_0^{s_t} P\,\mathrm{d}s$	$P\dfrac{s_1}{2}$

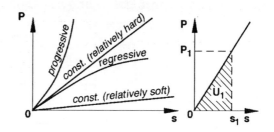

Springs in Tension and Compression

e.g., ring spring (Belleville spring)

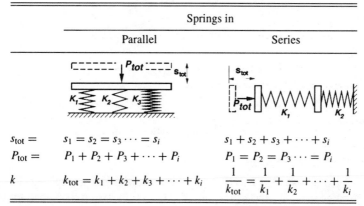

	Springs in	
	Parallel	Series
$s_{\text{tot}} =$	$s_1 = s_2 = s_3 \cdots = s_i$	$s_1 + s_2 + s_3 + \cdots + s_i$
$P_{\text{tot}} =$	$P_1 + P_2 + P_3 + \cdots + P_i$	$P_1 = P_2 = P_3 \cdots = P_i$
k	$k_{\text{tot}} = k_1 + k_2 + k_3 + \cdots + k_i$	$\dfrac{1}{k_{\text{tot}}} = \dfrac{1}{k_1} + \dfrac{1}{k_2} + \cdots + \dfrac{1}{k_i}$

Springs, continued

Springs in Bending

Rectangular, Trapezoidal, Triangular Springs

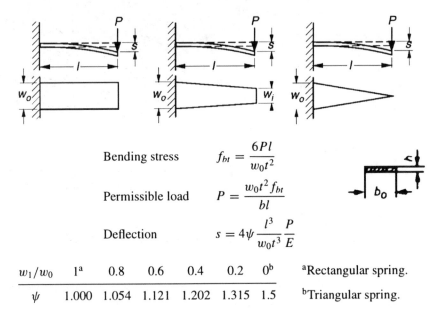

Bending stress $\qquad f_{bt} = \dfrac{6Pl}{w_0 t^2}$

Permissible load $\qquad P = \dfrac{w_0 t^2 f_{bt}}{bl}$

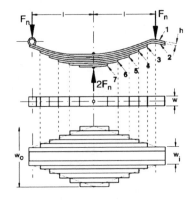

Deflection $\qquad s = 4\psi \dfrac{l^3}{w_0 t^3} \dfrac{P}{E}$

w_1/w_0	1[a]	0.8	0.6	0.4	0.2	0[b]	[a]Rectangular spring.
ψ	1.000	1.054	1.121	1.202	1.315	1.5	[b]Triangular spring.

$$f_{bt} = \text{permissible bending stress}$$

Laminated Leaf Springs

Laminated leaf springs can be imagined as trapezoidal springs cut into strips and rearranged (spring in sketch can be replaced by two trapezoidal springs in parallel) of total spring width

$$w_0 = zw_1$$

where z is the number of leaves.
Then

$$P_n \approx \dfrac{bt^2 f_{bt}}{6l}$$

If leaves 1 and 2 are the same length (as in the sketch),

$$w_1 = 2w$$

The calculation does not consider friction. In practice, friction increases the carrying capacity by between 2 and 12%.

Springs, continued

Disc Springs (Ring Springs)

Different characteristics can be obtained by combining n springs the same way and i springs the opposite way.

$$P_{tot} = n\,P_{single}$$

$$s_{tot} = i\,s_{single}$$

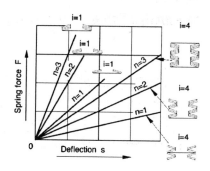

Material Properties

Hot-worked steels for springs to ASTM A 322, e.g., for leaf springs 9255; 6150 —resp. to BS 970/5, e.g., 250 A 53; 735 A 50—(Modulus of elasticity: $E = 200{,}000$ N/mm²).

f_{bt}: static: 910 N/mm²
oscillating: (500 ± 225) N/mm² (scale removed and tempered)

Coiled Torsion Spring

The type shown in the sketch has both ends free and must be mounted on a guide post. Positively located arms are better.

Spring force[a]	$P \approx \dfrac{Z f_{bt}}{r}$	
Angle of deflection	$\alpha \approx \dfrac{P r l}{I E}$	
Spring coil length (N = number of coils)	$l \approx D_m \pi N$	
Section modulus	$Z = \dfrac{I}{y}$	

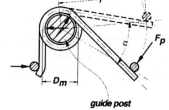

(Additional correction is needed for deflection of long arms.)

[a]Not allowing for the stress factor arising from the curvature of the wire.

Springs, continued

Springs in Torsion

Torsion Bar Spring

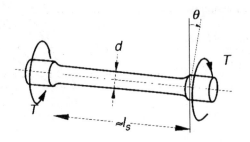

Shear stress	Torque	Angle of twist	
$\tau = \dfrac{5T}{d^3}$	$T = \dfrac{d^3}{5}\tau_{qt}$	$\vartheta = \dfrac{Tl_s}{GI_p}$	$\approx \dfrac{10Tl_s}{Gd^4}$

l_s = spring length as shown in sketch.

Stress τ_{qt} and fatigue strength τ_f in N/mm²

	Static		Oscillating[a]	
τ_{qt}	not preloaded	700	$\tau_f = \tau_m \pm \tau_A$	$d = 20$ mm $\quad 500 \pm 350$
	preloaded	1020		$d = 30$ mm $\quad 500 \pm 240$

[a] Surface ground and shot-blasted, preloaded.
τ_m = mean stress.
τ_A = alternating stress amplitude of fatigue strength.

Coefficients of Static and Sliding Friction[a]

Materials	Static		Sliding	
	Dry	Greasy	Dry	Greasy
Hard steel on hard steel	0.78	0.11	0.42	0.029
	—	—	—	0.12
Mild steel on mild steel	0.74	—	0.57	0.09
Hard steel on graphite	0.21	0.09		
Hard steel on babbitt (ASTM No. 1)	0.70	0.23	0.33	0.16
Hard steel on babbitt (ASTM No. 8)	0.42	0.17	0.35	0.14
Hard steel on babbitt (ASTM No. 10)	—	0.25	—	0.13
Mild steel on cadmium silver	—	—	—	0.097
Mild steel on phosphor bronze	—	—	0.34	0.173
Mild steel on copper lead	—	—	—	0.145
Mild steel on cast iron	—	0.183	0.23	0.133
Mild steel on lead	0.95	0.5	0.95	0.3
Nickel on mild steel	—	—	0.64	0.178
Aluminum on mild steel	0.61	—	0.47	
Magnesium on mild steel	—	—	0.42	
Magnesium on magnesium	0.6	0.08		
Teflon on Teflon	0.04	—	—	0.04
Teflon on steel	0.04	—	—	0.04
Tungsten carbide on tungsten carbide	0.2	0.12		
Tungsten carbide on steel	0.5	0.08		
Tungsten carbide on copper	0.35			
Tungsten carbide on iron	0.8			
Bonded carbide on copper	0.35			
Bonded carbide on iron	0.8			
Cadmium on mild steel	—	—	0.46	

(continued)

[a]Note that static friction between like materials (i.e., aluminum on aluminum, nickel on nickel, corrosion resistant steel on corrosion resistant steel) is high, and they often gall or sieze when used together dry.

Ref.: *Marks Standard Handbook for Mechanical Engineers*, 6th Edition, by T. Baumeister and E. Avallone. Copyright © 1958, McGraw–Hill, New York. Reproduced with permission of The McGraw–Hill Companies.

Coefficients of Static and Sliding Friction,[a] continued

Materials	Static		Sliding	
	Dry	Greasy	Dry	Greasy
Copper on mild steel	0.53	—	0.36	0.18
Nickel on nickel	1.10	—	0.53	0.12
Brass on mild steel	0.51	—	0.44	
Brass on cast iron	—	—	0.30	
Zinc on cast iron	0.85	—	0.21	
Magnesium on cast iron	—	—	0.25	
Copper on cast iron	1.05	—	0.29	
Tin on cast iron	—	—	0.32	
Lead on cast iron	—	—	0.43	
Aluminum on aluminum	1.05	—	1.4	
Glass on glass	0.94	0.01	0.40	0.09
Carbon on glass	—	—	0.18	
Garnet on mild steel	—	—	0.39	
Glass on nickel	0.78	—	0.56	
Copper on glass	0.68	—	0.53	
Cast iron on cast iron	1.10	—	0.15	0.070
Bronze on cast iron	—	—	0.22	0.077
Oak on oak (parallel to grain)	0.62	—	0.48	0.164
				0.067
Oak on oak (perpendicular)	0.54	—	0.32	0.072
Leather on oak (parallel)	0.61	—	0.52	
Cast iron on oak	—	—	0.49	0.075
Leather on cast iron	—	—	0.56	0.36
Laminated plastic on steel	—	—	0.35	0.05
Fluted rubber bearing on steel	—	—	—	0.05

[a]Note that static friction between like materials (i.e., aluminum on aluminum, nickel on nickel, corrosion resistant steel on corrosion resistant steel) is high, and they often gall or sieze when used together dry.

Ref.: *Marks Standard Handbook for Mechanical Engineers*, 6th Edition, by T. Baumeister and E. Avallone. Copyright © 1958, McGraw–Hill, New York. Reproduced with permission of The McGraw–Hill Companies.

Gears

Basic Formulas for Involute Gears

To obtain	Symbol	Spur gears	Helical gears
Pitch diameter	D	$D = \dfrac{N}{P}$	$D = \dfrac{N}{P_n \cos \psi}$
Circular pitch	p	$p = \dfrac{\pi}{P} = \dfrac{\pi D}{N}$	$p_n = \dfrac{\pi}{P_n}$
			$p_t = \dfrac{\pi D}{N} = \dfrac{p_n}{\cos \psi}$
			$p_x = \dfrac{1}{N} = \dfrac{p_n}{\sin \psi}$
Diametral pitch	P	$P = \dfrac{\pi}{p} = \dfrac{N}{D}$	$P_n = \dfrac{N}{D \cos \psi}$
Number of teeth	N	$N = PD = \dfrac{\pi D}{p}$	$N = P_n \cos \psi\, D$
Outside diameter	D_o	$D_o = D + 2a = \dfrac{N + 2}{P}$	$D_o = D + 2a$
Root diameter	D_r	$D_r = D_o - 2h_t$	$D_r = D_o - 2h_t$
Base diameter	D_b	$D_b = D \cos \phi$	$D_b = D \cos \phi_t$
Base pitch	p_b	$p_b = p \cos \phi$	$p_b = p_t \cos \phi_t$
Circular tooth thickness at D	t	$t = 0.5p = \dfrac{\pi D}{2N}$	$t_t = 0.5p_t$
			$t_n = 0.5p_n$

Ref.: *Marks Standard Handbook for Mechanical Engineers*, 6th Edition, by T. Baumeister and E. Avallone. Copyright © 1958, McGraw–Hill, New York. Reproduced with permission of The McGraw–Hill Companies.

Gears, continued

Bevel Gears

a	= standard center distance
b	= tooth width
h_{aO}	= addendum of cutting tool
h_{aP}	= addendum of reference profile
h_{fP}	= dedendum of reference profile
k	= change of addendum factor
p_e	= normal pitch ($p_e = p \cos \alpha$, $p_{et} = p_t \cos a_t$)
z	= number of teeth
z_{nx}	= equivalent number of teeth
(C_{perm}), C	= (permissible) load coefficient
F_t	= peripheral force on pitch cylinder (plane section)
K_I	= operating factor (external shock)
K_V	= dynamic factor (internal shock)
$K_{F\alpha}$	= end load distribution factor
$K_{F\beta}$	= face load distribution factor $\Big\}$ for root stress
K_{FX}	= size factor
$K_{H\alpha}$	= end load distribution factor
$K_{H\beta}$	= face load distribution $\Big\}$ for flank stress
R_e	= total pitch cone length (bevel gears)
R_m	= mean pitch cone length (bevel gears)
T	= torque
Y_F, (Y_S)	= form factor, (stress concentration factor)
Y_β	= skew factor
Y_ε	= load proportion factor
Z_H	= flank form factor
Z_ε	= engagement factor
Z_R	= roughness factor
Z_V	= velocity factor
α_P	= reference profile angle (DIN 867: $\alpha P = 20°$)
α_W	= operating angle
β	skew angle for helical gears, pitch cylinder
β_b	base cylinder
ρ	= sliding friction angle (tan $\rho = \mu$)
ρ_{a0}	= tip edge radius of tool
$\sigma_{F\,\text{lim}}$	= fatigue strength
$\sigma_{H\,\text{lim}}$	= Hertz pressure (contact pressure)

Gears, continued

Bevel Gears, Geometry

Cone angle δ:

$$\tan \delta_1 = \frac{\sin \Sigma}{\cos \Sigma + u}$$

$$\left(\Sigma = 90° \Rightarrow \tan \delta_1 = \frac{1}{u} \right)$$

$$\tan \delta_2 = \frac{\sin \Sigma}{\cos \Sigma + 1/u}$$

$$(\Sigma = 90° \Rightarrow \tan \delta_2 = u)$$

$$\left. \begin{array}{l} \text{angle between} \\ \text{shafts} \end{array} \right\} \Sigma = \delta_1 + \delta_2$$

$$\left. \begin{array}{l} \text{external pitch} \\ \text{cone distance} \end{array} \right\} R_e = \frac{d_e}{2 \sin \delta}$$

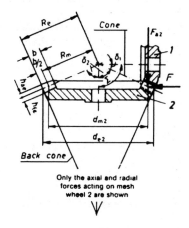

Only the axial and radial
forces acting on mesh
wheel 2 are shown

Development of the back cone to examine the meshing conditions gives the virtual cylindrical gear (suffix "v" = virtual) with the values

$$\left. \begin{array}{l} \text{straight} \\ \hline \text{spiral} \end{array} \right| \text{bevel gears} \left| \begin{array}{l} z_v = \dfrac{z}{\cos \delta} \\[2mm] z_v \approx \dfrac{z}{\cos \delta \times \cos^3 \beta} \end{array} \right| u_v = \dfrac{z_{v2}}{z_{v1}}$$

Bevel Gears, Design

The design is referred to the midpoint of the width b (suffix "m") with the values

$$R_m = R_e - \frac{b}{2} \qquad \bigg| \qquad m_m = \frac{d_m}{z}$$

$$d_m = 2 R_m \sin \delta \qquad \bigg| \qquad F_{tm} = \frac{2T}{d_m}$$

Axial and Radial Forces in Mesh

axial force $\quad F_a = F_{tm} \tan \alpha_n \times \sin \delta$

radial force $\quad F_r = F_{tm} \tan \alpha_n \times \cos \delta$

Gears, continued

Load Capacity of Tooth Root (Approximate Calculation)

Safety factor S_F against fatigue failure of tooth root:

$$S_F = \frac{\sigma_{F\,lim}}{\frac{F_{tm}}{bm_{nm}} \times Y_F \times Y_{EV} \times Y_\beta} \times \frac{Y_S \times K_{FX}}{K_I \times K_V \times K_{F\alpha} \times K_{F\beta}} \geq S_{F\,min}$$

Giving the approximate formula

$$m_{nm} \geq \frac{F_{tm}}{b} \times Y_F \times K_I \times K_V \times \underbrace{Y_{EV} \times Y_\beta \times K_{F\alpha}}_{\approx 1} \times \underbrace{\frac{K_{F\beta}}{Y_S \times K_{FX}}}_{\approx 1} \times \frac{S_{F\,min}}{\delta_{F\,lim}}$$

Y_F: substitute the number of teeth of the complementary spur gear z_v or, with spiral gears, $z_{vn} \approx z_v / \cos^3 \beta$.

Load Capacity of Tooth Flank (Approximate Calculation)

Safety factor S_H against pitting of tooth surface.

$$S_H = \frac{\sigma_{H\,lim}}{\sqrt{\frac{u+1}{u} \times \frac{F_{tm}}{b\,d_1}} \times Z_H \times Z_M \times Z_{EV}} \times \frac{Z_V \times K_{HX} \times Z_R \times K_L}{\sqrt{K_I \times K_V \times K_{H\alpha} \times K_{H\beta}}} \geq S_{H\,min}$$

For metals the material factor Z_M is simplified to

$$Z_M = \sqrt{0.35E} \quad \text{with} \quad E = \frac{2\,E_1\,E_2}{E_1 + E_2}$$

Giving the approximate formula

$$d_{vm1} \geq \sqrt{\frac{2T_1}{b} \times \frac{u_v + 1}{u_v} 0.35E} \times Z_{HV} \times \underbrace{Z_{EV} \times \sqrt{K_{H\alpha}}}_{\approx 1}$$

$$\times \frac{\sqrt{K_I \times K_V} \times \sqrt{K_{H\beta}}}{Z_V \times K_{HX} \times Z_R \times K_L} \times \frac{S_{H\,min}}{\sigma_{H\,lim}}$$

Gears, continued

Worm Gearing

Worm Gearing, Geometry

(Cylindrical worm gearing, normal module in axial section, BS 2519, angle between shafts $\Sigma = 90°$.)

Drive Worm

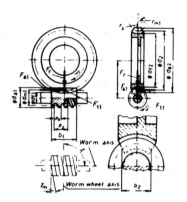

All the forces acting on the teeth in mesh are shown by the three arrows F_a, F_t, and F_r.
In the example,

$$z_1 = 2, \text{ right-hand helix}$$

	Worm, suffix 1	Worm wheel, suffix 2
Module	$m_x = m = m_t$	
Pitch	$p_x = m\,\pi = p_2 = d_2\pi/z_2$	
Mean diameter	$d_{m1} = 2\,r_{m1}$	
(free to choose, for normal values see DIN 3976)		
Form factor	$q = d_{m1}/m$	
Center helix angle	$\tan\gamma_m = \dfrac{mz_1}{d_{m1}} = \dfrac{z_1}{q}$	
Pitch diameter		$d_2 = mz_2$
Addendum	$h_{a1} = m$	$h_{a2} = m(1+x)^{\mathrm{a}}$
Dedendum	$h_{f1} = m(1 + c_1^*)$	$h_{f2} = m(1 - x + c_2^*)$
Tip clearance factor	$c_1^* = (0.167\ldots\underline{0.2}\ldots0.3) = c_2^*$	
Outside diameter	$d_{a1} = d_{m1} + 2h_{a1}$	$d_{a2} = d_2 + 2h_{a2}$
Tip groove radius		$r_k = a - d_{a2}/2$
Tooth width	$b_1 \geq \sqrt{d_{a2}^2 - d_2^2}$	$b_2 \approx 0.9\,d_{m1} - 2m$
Root diameter	$d_{f1} = d_{m1} - 2h_{f1}$	$d_{f2} = d_2 - h_{f2}$
Center distance	$a = (d_{m1} + d_2)/2 + xm$	

[a] Profile offset factor x for check of a preset center distance. Otherwise, $x = 0$.

Gears, continued

Worm Gearing, Design (Worm Driving)

	Worm	Worm wheel
Peripheral force	$F_{t1} = \dfrac{2\,T_1}{d_{m1}} K_I \times K_V$	$F_{t2} = F_{a1}$
Axial force	$F_{a1} = F_{t1} \times \dfrac{1}{\tan(\gamma + \rho)}$	$F_{a2} = F_{t1}$
Radial force	$F_{r1} = F_{t1} \times \dfrac{\cos\rho \times \tan a_n}{\sin(\gamma + \rho)}$	$= F_r = F_{r2}$
Rubbing speed	$v_g = \dfrac{d_{m1}}{2} \times \dfrac{\omega_1}{\cos\gamma_m}$	

Efficiency

Worm driving	Worm wheel driving
$\eta = \tan\gamma_m / \tan(\gamma_m + \rho)$	$\eta' = \tan(\gamma_m - \rho)/\tan\gamma_m$
	$(\gamma_m < \rho) \Rightarrow$ self-locking

Coefficient of Friction (Typical Values) $\mu = \tan\rho$

	$v_g \approx 1$ m/s	$v_g \approx 10$ m/s
Worm teeth hardened and ground	0.04	0.02
Worm teeth tempered and machine cut	0.08	0.05

Gears, continued

Calculation of Modulus m

Load capacity of teeth root and flanks and temperature rise are combined in the approximate formula

$$F_{t2} = C\, b_2\, p_2 \quad \text{where } b_2 \approx 0.8\, d_{m1} \text{ and } p_2 = m\pi$$

$$m \approx \sqrt[3]{\frac{0.8 T_2}{C_{\text{perm}\, q\, z_2}}} \quad \left| \begin{array}{l} F_{t2} = 2T_2/d_2 = 2T_2/(m\, z_2) \\ q \approx 10 \text{ for } i = 10, 20, 40 \\ q \approx 17 \text{ for } i = 80, \text{ self-locking} \end{array} \right.$$

Assumed values for normal, naturally cooled worm gears (worm hardened and ground steel, worm, wheel of bronze)

v_g	m s^{-1}	1	2	5	10	15	20
C_{perm}	N mm^{-2}	8	8	5	3.5	2.4	2.2

When cooling is adequate this value can be used for all speeds

$$C_{\text{perm}} \geq 8\,\text{N mm}^{-2}$$

Epicyclic Gearing

Velocity diagram and angular velocities
(referred to fixed space)

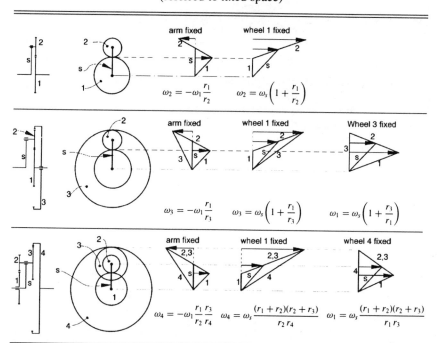

Formulas for Brakes, Clutches, and Couplings

The kinetic energy E of a rotating body is

$$E = \frac{I_g \omega^2}{2} = \frac{Wk^2}{g} \times \frac{N^2}{182.5} = \frac{Wk^2 N^2}{5878} \text{ ft lb}$$

where

I_g = mass moment of inertia of the body, lb ft s^2
k = radius of gyration of the body, ft^2
W = weight of the body, lb
N = rotational speed, rpm
g = gravitational acceleration, ft/s^2

If the angular velocity, ω rad/s, of the body changes by ΔN rpm in t seconds, the angular acceleration is

$$\alpha = \frac{2\pi \Delta N}{60t} = \frac{\Delta \omega}{t}$$

The torque T necessary to impart this angular acceleration to the body is

$$T = I_g \alpha = \frac{Wk^2}{32.2} \times \frac{2\pi \Delta N}{60t} = \frac{Wk^2 \Delta N}{308t} \text{ ft lb}$$

When a torque source drives a load inertia through a gear train, the equivalent inertia of the load must be used when calculating the torque required to accelerate the load.

$$I_{\text{equivalent}} = \left(\frac{N_2}{N_1}\right)^2 I_{\text{load}}$$

where

$I_{\text{equivalent}}$ = equivalent moment of inertia
I_2 = moment of inertia of the load
N_1 = rotational speed of the torque source
N_2 = rotational speed of the load

For example, in the following figure, the equivalent inertia seen by the clutch is

$$I_{\text{clutch}} = I_1 + I_2 \left(\frac{N_2}{N_1}\right)^2 + I_3 \left(\frac{N_3}{N_1}\right)^2$$

Formulas for Brakes, Clutches, and Couplings, continued

where

I_1 = moment of inertia of clutch and attached shaft 1 and gear
I_2 = moment of inertia of shaft 2 and attached gears
I_3 = moment of inertia of load and attached shaft and gear
N_1 = rotational speed of the clutch and attached shaft and gear
N_2 = rotational speed of shaft 2 and attached gears
N_3 = rotational speed of the load and attached shaft 3 and gear

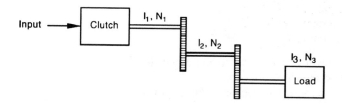

Work is the product of the magnitude of a force and the distance moved in the direction of the force. Power is the time rate at which the work is performed. The English unit of power is

$$1\,HP = 550\,ft\,lb/s = 33,000\,ft\,lb/min$$

Hence,

$$HP = \frac{P2r\pi N}{33,000} = \frac{PrN}{5252} = \frac{TN}{5252} \cdots$$

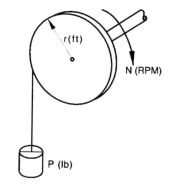

Pumps

Pump Relationships

b.hp. = brake horsepower, hp
D = impeller diameter, in.
f.hp. = fluid horsepower, hp
g = gravitational acceleration
H_p = fluid static head, ft
H_{sv} = suction head above vapor pressure, ft
H_t = fluid total head, ft
H_v = fluid velocity head, ft
H_{vp} = fluid vapor pressure head, ft
N = rotational speed, rpm
N_s = pump specific speed, (rpm $\sqrt{\text{gpm}}$)/ft$^{3/4}$
Q = volume flow rate, gpm
S = suction specific speed, (rpm $\sqrt{\text{gpm}}$)/ft$^{3/4}$
V = fluid velocity, ft/s
δ = fluid specific gravity
η = overall efficiency, %
Φ = overall head rise coefficient at point of maximum efficiency

Pump specific speed:

$$N_s = N\frac{\sqrt{Q}}{H^{3/4}}$$

Fluid velocity head:

$$H_v = \frac{V^2}{2g}$$

Fluid total head:

$$H_t = H_p + H_v$$

Suction head above vapor pressure:

$$H_{sv} = H_p + H_v - H_{vp}$$

Suction specific speed:

$$S = \frac{N\sqrt{Q}}{H_{sv}^{3/4}}$$

Overall efficiency:

$$\eta = \frac{\text{f.hp.}}{\text{b.hp.}}$$

Fluid horsepower:

$$\text{f.hp.} = -\frac{QH\delta}{3960}$$

Impeller diameter:

$$D = \frac{1840\Phi\sqrt{H}}{N}$$

Pumps, continued

Approximate Relative Impeller Shapes and Efficiencies as Related to Specific Speed

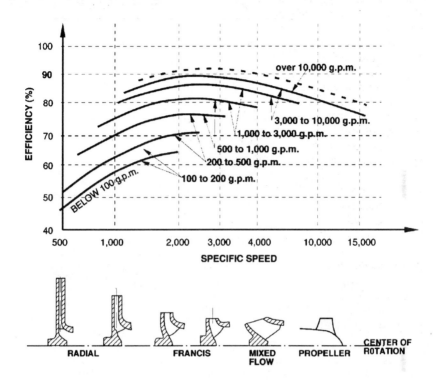

MECHANICAL DESIGN

Control System Loads

Control system	Type	Ult. design load (1.5 factor of safety included[a]), lb
Elevator	Stick	450/375
	Wheel	450
Aileron	Stick	150
	Wheel	240
Rudder & brake	Pedal	450
Flap, tab, stabilizer, spoiler, landing gear, arresting hook, wing-fold controls	Crank, wheel, or lever operated by push or pull	$\left(\dfrac{1+R}{3}\right)^{b}(50)(1.5)^{c,d}$
	Small wheel or knob	200 in.-lb[c,e] 150[b,f]

[a] For dual systems, design for 75% of two-pilot control loads from control bus to control surface connection.

[b] R = Radius of wheel or length of lever.

[c] Applied at circumference of wheel, or grip of crank, or lever, and allowed to be active at any angle within 20 deg of plane of control.

[d] But not less than 75 lb or more than 225 lb.

[e] If operated only by twist.

[f] If operated by push or pull.

Ref.: *Federal Aviation Regulation* (FAR) Part 25 and MIL-A-008865A.

Dynamic-Stop Loads

Loads due to stopping a moving mass can be estimated from the following:

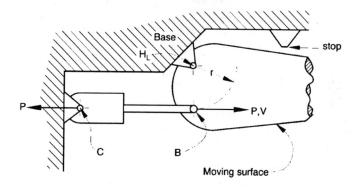

where

B = actuator to surface pivot
C = actuator to base pivot
H_L = surface hinge line
P = dynamic-stop load
k = linear spring rate (lb/in.) includes actuator and moving surface
m^a = equivalent mass at $B = I_p/386r^2$
V = linear velocity of B in direction of stop reaction, in./s
I_p = polar moment of inertia of mass about H_L, lb-in.²
    [a]for linear movement; $m = $ (weight/386)

Friction of Pulleys and Rollers

A study of the following formula shows friction moment (M_F) will be a minimum when the diameter of the bearing is as small as practical relative to the diameter of the pulley.

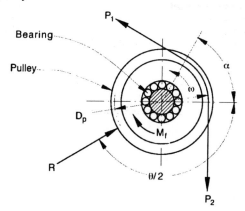

Friction of Pulleys and Rollers, continued

$$M_F = \frac{R\mu_{br}d_{br}}{2}$$

$$= \frac{(P_1 - P_2)D_p}{2}$$

$$R = (P_1 + P_2)\sin\frac{\theta}{2}$$

α = wrap angle

d_{br} = bearing diameter

$\theta = 2\pi - \alpha$

μ_{br} = bearing coefficient of friction

ω = relative motion

If the bearing is frozen, friction of the cable on the pulley governs. Then,

$$P_1 = P_2(e^{\mu\alpha}) = P_2(10^{0.4343\mu\alpha})$$

where

μ = coefficient of friction of cable to pulley
α = angle of wrap, rad

Coefficients of Friction

The following values are recommended for Teflon-lined lubricated metal, plain spherical bearings, and sliding surfaces.

Material in contact with polished or chromed steel	μ
Teflon-lined	0.10
Alum. bronze, beryllium copper	0.15
Steel	0.30
Dry nonlubricated bushings or bearings	≤ 0.60
Ball and roller bearings in the unjammed or noncorroded condition	0.01

Cables

Aircraft cable is an efficient means for transmitting control loads over long distances. To ensure that a control loop will carry load in both legs and to reduce cable load deformation, preloading or rigging load is applied.

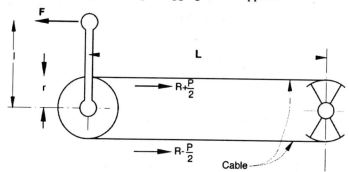

where

$d = \frac{PL}{2A_c E_c} \quad R - \frac{P}{2} > 0$

$P = F(\ell/r)$, active cable load

d = cable extension or contraction between supports due to load application

R = rigging load

A_c = cable cross-sectional area

E_c = cable modulus of elasticity

As shown, the rigging load must be greater than one-half of the active load to reduce cable deflection and ensure that the cable will not become slack and cause a different load distribution.

Airframe Deformation Loads

Usually the cable run cannot be located at the airframe structure neutral axis, and so rigging loads will be affected by structure deformations.

An estimate of the change in rigging load ΔR due to structure deformation is

$$\Delta R \approx \frac{f}{E_s} A_c E_c$$

where

f = average airframe material working stress along the line of cable supports (if compression, P will be a negative load)

E_s = modulus of elasticity of structure

Rigging loads are also affected by temperature changes when steel cables are used with aluminum airframes.

Cables, continued

Cable-Sheave Friction with Groove Radius = (d/2)

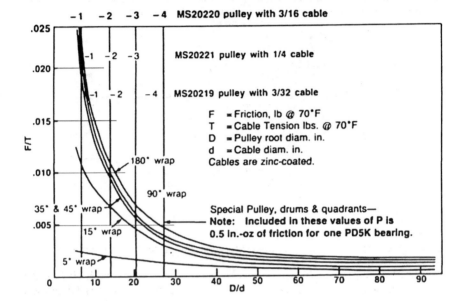

Cable-Sheave Friction with Groove Radius = (d/2) + 0.02

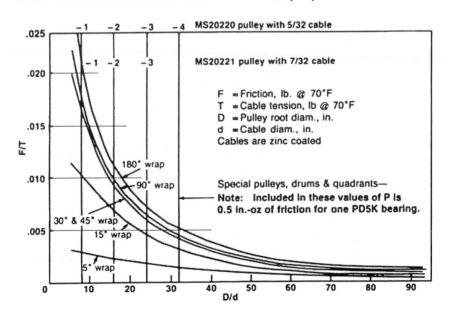

Bearings and Bushings

The radial limit load P_s for ball bearings may be obtained from the following expression:

$$P_s = knD^2 \text{ lb}$$

where

k = a design factor (see table)
n = number of balls
D = ball diameter (in.)

Bearing type	Design k factors
Deep groove	10,000
Single row, self-aligning	4,800
Double row, self-aligning	3,800
Rod end	3,200

The basic dynamic capacity is the constant radial load at which 10% of the bearings tested fail within 2000 cycles. (For rating purposes a cycle is defined as a 90-deg rotation from a fixed point and return.) The dynamic load capacity P_d for N cycles is given by:

$$P_d = \frac{D_0}{L}$$

where D_0 = basic dynamic capacity and

$$L = \left(\frac{N}{2000}\right)^{1/3.6} \qquad \text{life factor}$$

Variable Dynamic Loads

If the dynamic loads on a bearing vary greatly, the following equation may be used to estimate the representative design load

$$P_e = \sqrt[3.6]{\frac{\Sigma N(P)^{3.6}}{\Sigma N}}$$

where

P_e = equivalent dynamic design load to give the same life as the variable loads
N = number of revolutions for a particular value of P
P = load acting for a particular value of N

Bearings and Bushings, continued

Combined Loads

For combined radial and thrust loading, either static or dynamic, the equivalent radial load P_{e_r} is

$$P_{e_r} = R + YT$$

where

$R =$ applied radial load
$Y =$ (radial load rating/thrust load rating)
$T =$ applied thrust load

Contact Stresses

Contact stress due to spheres or cylinders on various surfaces can be calculated from the following expressions where

E = modulus of elasticity
μ = Poisson's ratio
R = reaction
D = diameter of cylinder or sphere
D_b = diameter of base
$K, k =$ constant dependent on material and shape of the race to be found by references not in this handbook

Compressive Stress on Contact Point

$$f_{c_{(cyl)}} = k\sqrt{\frac{pE}{D}}$$

where $p =$ load/in.

$$f_{c_{(sph)}} = K\sqrt[3]{P\left(\frac{E}{D}\right)^2}$$

where $P =$ total load.

Maximum Tensile Stress in Region of Contact Point

$$\tau_{max} = -f_c\left(\frac{1 - 2\mu}{3}\right)$$

Maximum Shear Stress in Region of Contact Point

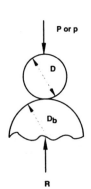

P or p

D

D_b

R

$$f_{s_{max}} = 0.33 f_c$$

(Spheres or end view of cylinders)—convex case shown.

Bearings and Bushings, continued

General Bearing Characteristics

Operating parameter	Rolling element	Sliding surface
Load		
static	Low (ball) to medium (needle)	High
oscillating	Low	High
rotating	High	Low
dynamic	Poor	Good
vibration & impact	Poor	Good
Life	____[a]	____[b]
Speed	High	Low
Friction	Low	High
Noise	____[c]	Low
Damping	Poor	Medium
Envelope restrictions		
radial	Large	Small
axial	Small	Small
Lubricant type	Oil or grease	Grease, solid dry film, or none
Cost	Medium	Low
Type of failure	____[d]	____[e]
Power requirement	Low	High

[a]Limited by properties of bearing metal, lubrication, and seals.
[b]Limited by resistance to wear, galling, fretting, seizing, and lubrication.
[c]Depends on bearing quality and mounting.
[d]Quite rapid, with operations severely impaired.
[e]Gradual wear, bearing operable unless seizure occurs.

Bearings and Bushings, continued

Friction Calculation for Airframe Bearings

The total running friction torque in an airframe bearing may be estimated by means of the following equation:

$$F_t \approx \frac{ST}{12} + \frac{PD}{1000}$$

where

D = OD of bearing, in.
F_t = total running friction torque, in.-lb
S = speed factor from Bearing Friction Speed Factor chart
T = torque (in.-oz) from the Airframe Bearings Friction Torque chart and table
P = applied radial load, lb
r_b = bore radius, in.

1000 is an empirical factor based on test results.

Bearing Friction Speed Factor

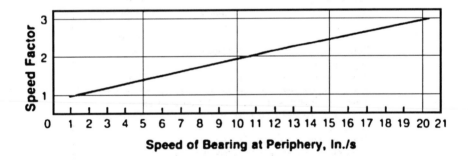

Bearings and Bushings, continued

Airframe Bearings Friction Torque

For maximum expected torque, multiply values from curve by 1.50.

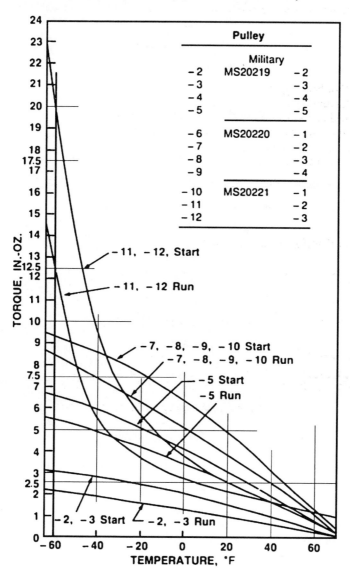

Bearings and Bushings, continued

Machinery Bearings Friction Torque

The friction of rolling element bearings will vary with the type of lubrication, temperature, bearing material, and bearing type. The following formula and table can be used to estimate friction torque.

$$F_t \approx \mu r_b P$$

Values of μ are given in the following table.

Friction coefficients applicable to rolling element machinery bearings

Bearing type	Friction coefficient,[a] μ
Self-aligning ball	0.0010
Cylindrical roller	0.0011
Thrust ball	0.0013
Single-row ball	0.0015
Tapered roller	0.0018
Full-complement needle	0.0025

[a]Friction coefficient referred to bearing bore radius.

Rods and Links

Rods and links used as tension members are designed for the most severe combination of loads. These include rigging, structure deformation, temperature, and any beam effect loads applied in addition to operation loads. Any rod or link that it is possible to grasp or step on at any time during construction, maintenance, or inspection should be checked independently for a 150-lb ultimate lateral load applied at any point along its length. Transverse vibration and fatigue checks should also be made on all control rods.

Rods and links used as compression members should be avoided if possible. When they are necessary, a beam column analysis should be made. Buckling due to column or transverse load conditions is the most critical mode of failure. The stiffness of a link support will also affect the stability of the link under compression. The equation presented below enables an instability load to be determined based on existing spring rates at each of the link ends. In a system consisting of several links, it is necessary to check for the instability of the whole system.

Rods and Links, continued

Link Stability

Critical buckling load P_{cr} for the link is

$$P_{cr} = \frac{k_1 k_2 L}{k_1 + k_2}\left[1 - \left(\frac{e}{L}\right)^{2/3}\right]^{3/2}$$

where

k_1, k_2 = spring rates in lb/in.
e = initial eccentricity
L = link length

Temperature Loads

Change in temperature that will cause buckling in rods or links is

$$\Delta T_{cr} = \left(C\pi^2 E_t l - P_a L^2\right)/E_s \alpha A L^2$$

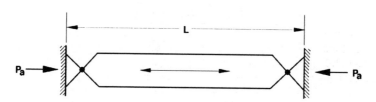

where

α = coefficient of thermal expansion
C = end restraint constant
E_t = tangent modulus
E_s = secant modulus
P_a = applied mechanical load

Rods and Links, continued

Change in tension in a rod, link, or cable due to a change in temperature is

$$\Delta P_t = \Delta T(\alpha_1 - \alpha_2) \Big/ \left(\frac{1}{A_1 E_1} + \frac{1}{A_2 E_2} \right)$$

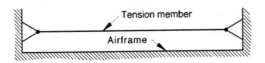

where

ΔT = change in temperature

α_1, α_2 = coefficient of linear expansion of airframe and member, respectively

$A_1 E_1, A_2 E_2$ = cross-sectional area times modulus of elasticity of airframe and member, respectively

A_1 (effective) for the airframe is usually quite large compared to A_2; therefore, in many cases, $1/A_1 E_1$ can be neglected without serious error, and $\Delta P_t \approx \Delta T(\alpha_1 - \alpha_2) A_2 E_2$.

Shafts

Axles and Shafts (Approximate Calculation)

Axles

Axle type	Required section modulus for bending	Solid axle of circular cross section ($Z \approx d^3/10$)	Permissible bending stress[a]
fixed[b]			$f_{bt} = \dfrac{f_{bt\,U}}{(3\ldots5)}$ [c]
	$Z = \dfrac{M}{f_{bt}}$	$d = \sqrt[3]{\dfrac{10M}{f_{bt}}}$	
rotating			$f_{bt} = \dfrac{f_{bt\,A}}{(3\ldots5)}$ [c]

[a] f_{bt} allows for stress concentration-, roughness-, size-, safety-factor and combined stresses—see Gieck reference.
[b] Formulas are restricted to load classes I and II.
[c] $3 \ldots 5$ means a number between 3 and 5.

Shafts, continued

Shafts

Stress	diameter for solid shaft	permissible torsional stress[a]
pure torsion		$\tau_{qt} = \dfrac{\tau_t U}{(3\ldots5)}$ [b]
	$d = \sqrt[3]{\dfrac{5T}{\tau_{qt}}}$	
torsion and bending		$\tau_{qt} = \dfrac{\tau_t U}{(10\ldots15)}$ [c]

[a] τ_{qt} allows for stress concentration-, roughness-, size-, safety-factor and combined stresses—see Gieck reference.
[b] $3\ldots5$ means a number between 3 and 5.
[c] $10\ldots15$ means a number between 10 and 15.

Bearing Stress

$$\left.\begin{array}{l}\text{on shaft}\\ \text{extension}\end{array}\right\} f_{bm} = \frac{P}{db} \le f_b$$

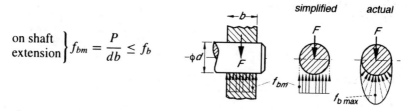

Shear Due to Lateral Load

Calculation unnecessary when

		with		cross
$l > d/4$	for all shafts		circular	
$l > 0.325\,h$	for fixed axles		rectangular	sections

where

l = moment arm of force F
M, T = bending moment, torque, respectively
$f_{bm}, (f_b)$ = mean, (permissible) bearing stress

Pins

For hollow pins, keep D_0/t at 7.0 or below to avoid local pin deformation. Check pins for shear and bending as shown for double shear symmetrical joints. Shear stress:

$$\tau_{au} = \frac{P}{1.571(D_0^2 - D_1^2)}$$

Bending stress:

$$f_{bu} = \frac{M_{max}}{0.0982k_b(D_0^3 - D_1^3)}$$

where

k_b = plastic bending coefficient (1.0 to 1.7) (usually 1.56 for pins made from material with more than 5% elongation)

$$M_{max} = \frac{P}{2}\left(\frac{t_1}{2} + \frac{t_2}{4} + g\right)$$

Pin clamp-up stress will be additive to bending and tension stress.

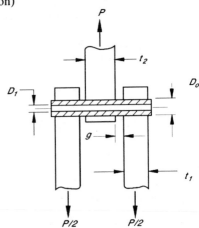

Actuators

In general, hydraulic or pneumatic actuators are strength checked as pressure vessels—with inherent discontinuities and stress risers—and as stepped columns. Jackscrew actuators are also checked as columns. In jackscrews, combined compression and torque stresses at the thread root along with the reduced section make the effective length longer than the physical support length.

The system loaded by actuators must withstand maximum actuator output load. Design pressures for hydraulic and pneumatic actuators are defined in MIL-H-5490 and MIL-P-5518. Those pressures are defined as a percentage of nominal system pressure as shown in the following table.

Actuators, continued

	Load factor, %	
Pressure	Hydraulic	Pneumatic
Operating	100	100
Proof	150[a]	200
Burst	300	400

[a]Factors apply to nominal pressure except the
150 applies to relief valve setting (usually 120%
of nominal).

Cylindrical bodies are subject to hoop and axial tension due to internal pressure. Minimum thickness of the cylinder wall is

$$t_c = \frac{pD}{2f_{tU}}$$

where

p = burst pressure
D = inside diameter of the cylinder
f_{tU} = ultimate tensile strength

t_c should be such that $d/t_c < 30$ to preclude excessive deflection and piston seal leakage.

The actuator piston rod is subject to an external pressure. The required wall thickness of a hollow piston rod is

$$t_p = D\sqrt[3]{\frac{p(1 - \mu)^2}{2E}}$$

where

p = burst pressure
D = piston rod outside diameter
μ = Poisson's ratio (0.3 for metals)

Stress concentrations at cylinder ends, ports, and section changes should be carefully considered and stress kept at a low enough level to meet fatigue requirements.

Actuators, continued

Margin of Safety

To determine whether an actuator will fail under the combined action of the applied compressive load, end moment, and moment due to eccentricities, the following two margins of safety must be obtained. To preclude failure, the resulting margins of safety must be greater than or equal to zero.

Piston rod with bending and column action:

$$MS = \frac{1}{R_c + R_b} - 1$$

where

$$R_c = \frac{P}{P_{cr}} \qquad R_b = \frac{f_b}{f_{bU}}$$

f_b = bending stress including beam column effect
f_{bU} = bending modulus of rupture
P = applied comprehensive load
P_{cr} = column critical load

Cylinder with combined bending and longitudinal and hoop tension:

$$MS = \frac{1}{\sqrt{R_b^2 + R_{gt}^2 + R_b R_{ht}}} - 1$$

where

$$R_b = \frac{f_b}{f_{by}} \qquad R_{ht} = \frac{f_{htU}}{f_{ty}}$$

f_{ty} = tensile yield stress
f_{htU} = applied ultimate hoop tension stress
f_{by} = bending modulus of yield
f_b = applied bending stress including beam column effect

Cylindrical Fits

Tolerances and allowances for cylindrical fits are given in the following tables. Hole diameter shall be selected in accordance with standard reamer sizes and called out on drawings as decimal conversions. Take tolerances for hole diameter from the Press, Sliding, and Running Fits tables.

Always specify surface finish on drawings for press, sliding, and running fits.

Classification of Fits

Press Fit:

Force = Applied by hydraulic press for units permanently assembled
Drive = Applied by arbor press for replaceable or removable units

Sliding Fit:

Normal = Shafts and gears, clutches, or similar parts that must be free to slide
Close = Intended for the accurate locations of parts that must assemble without perceptible play, such as a bearing in a gear housing

Running Fit:

Easy = Mechanisms easily assembled (high-speed parts)
Close = Link mechanisms with close tolerances (low-speed parts)
Fine = Mechanisms where extreme accuracy and minimum clearance are desired (parts having a slow rotational motion)

Press Fits[a]

| Hole | | Force fit | | | | Drive fit | | | |
| | | Shaft | | Interference (reference) | | Shaft | | Interference (reference) | |
Basic diam from–thru	Tol.	Allow.	Tol.	Min.	Max.	Allow.	Tol.	Min.	Max.
0–0.49	+0.0005 −0.0000	+0.0015	+0.0000 −0.0005	0.0005	+0.0015	+0.0012	+0.0000 −0.0005	0.0002	0.0012
0.50–0.99	+0.0007 −0.0000	+0.002	+0.0000 −0.0005	0.0008	0.002	+0.0017	+0.0000 −0.0005	0.0005	0.0017

[a]Dimensions in inches. Ref.: USAS B4.1-1965. On cylindrical fits, the hole diameter should be held constant. Clearance or interference is created by varying the shaft diameter. Tolerance on the hole diameter is always positive. Tolerance on the shaft diameter is always negative.

Sliding Fits[a]

| Hole | | Normal fit | | | | Close fit | | | |
| | | Shaft | | Clearance (reference) | | Shaft | | Clearance (reference) | |
Basic diam from–thru	Tol.	Allow.	Tol.	Min.	Max.	Allow.	Tol.	Min.	Max.
0–0.49	+0.0005 −0.0000	−0.001	+0.000 −0.001	0.001	0.0025	−0.0005	+0.0000 −0.0005	0.0005	0.0015
0.50–0.99	+0.0007 −0.0000	−0.0015	+0.000 −0.001	0.0015	0.0032	−0.0005	+0.0000 −0.0005	0.0005	0.0017

[a]Dimensions in inches. Ref.: USAS B4.1-1965. On cylindrical fits, the hole diameter should be held constant. Clearance or interference is created by varying the shaft diameter. Tolerance on the hole diameter is always positive. Tolerance on the shaft diameter is always negative.

Running Fits[a]

Basic diam from–thru	Hole Tol.	Easy fit Shaft Allow.	Easy fit Shaft Tol.	Easy fit Clearance (reference) Min.	Easy fit Clearance (reference) Max.	Close fit Shaft Allow.	Close fit Shaft Tol.	Close fit Clearance (reference) Min.	Close fit Clearance (reference) Max.	Fine fit Shaft Allow.	Fine fit Shaft Tol.	Fine fit Clearance (reference) Min.	Fine fit Clearance (reference) Max.
0–0.49	+0.0005 −0.0000	−0.001	+0.0000 −0.0005	0.001	0.002	−0.0005	+0.0000 −0.0005	0.0005	0.0015	−0.0005	+0.0000 −0.0005	0.0005	0.0015
0.50–0.99	+0.0007 −0.0000	−0.001	+0.000 −0.001	0.001	0.0027	−0.0005	+0.0000 −0.0005	0.0005	0.0017	−0.0005	+0.0000 −0.0005	0.0005	0.0015

[a]Dimensions in inches. Ref.: USAS B4.1-1965. On cylindrical fits, the hole diameter should be held constant. Clearance or interference is created by varying the shaft diameter. Tolerance on the hole diameter is always positive. Tolerance on the shaft diameter is always negative.

Key Joints

Tapered Keys (Cotter Pins)

Safety margin for load P: In view of the additional load produced when the key is tightened, a 25% margin should be added in the following.

Contact Pressure Between

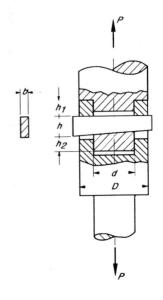

$$\text{rod and key } f = \frac{1.25\,P}{bd} \le f_{bU}$$

$$\text{socket and key } f = \frac{1.25\,P}{b(D-d)} \le f_{bU}$$

Required Key Width

$$h = 0.87\sqrt{\frac{0.625\,P(D+d)}{bf_{bU}}} \qquad h_1 = h_2 = k \qquad 0.5h < k < 0.7h$$

Shear Stress

$$\tau = \frac{1.5F}{bh} \le \tau_{qU}$$

Key Joints, continued

Load on Screw

Required root diameter d_k when load is applied

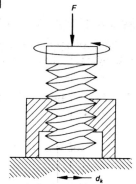

a) during screwing operation

$$d_k = \sqrt{\frac{4P}{\pi \times 0.75 f_U}}$$

b) after

$$d_k = \sqrt{\frac{4P}{\pi \times f_U}}$$

where P = load to be moved and f_{bU}, f_U, and τ_{qU} are allowable stresses.

Kinematics

Simple Connecting-Rod Mechanism

$$s = r(1 - \cos\varphi) + \frac{\lambda}{2} r \sin^2\varphi$$

$$v = \omega r \sin\varphi (1 + \lambda\cos\varphi)$$

$$a = \omega^2 r(\cos\varphi + \lambda\cos 2\varphi)$$

$$\lambda = \frac{r}{l} = \frac{1}{4} \text{ to } \frac{1}{6} \quad (\lambda \text{ is called the crank ratio})$$

$$\varphi = \omega t = 2\pi n t$$

Scotch-Yoke Mechanism

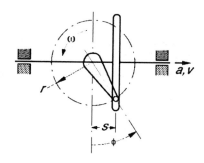

$$s = r\sin(\omega t)$$

$$v = \omega r\cos(\omega t)$$

$$a = -\omega^2 r\sin(\omega t)$$

$$\omega = 2\pi n$$

Kinematics, continued

Cardan Joint

For constant input speed, the output speed will be

Variable	Constant (due to auxiliary shaft H)

For all shafts located in one plane, the following relations apply.

$$\tan \varphi_2 = \tan \varphi_1 \times \cos \beta \qquad \tan \varphi_3 = \tan \varphi_1 \qquad\qquad \tan \varphi_3 = \tan \varphi_1$$

$$\omega_2 = \omega_1 \frac{\cos \beta}{1 - \sin^2 \beta \times \sin^2 \varphi_1} \qquad \omega_3 = \omega_1 \qquad\qquad \omega_3 = \omega_1$$

$$\alpha_2 = \omega_1^2 \frac{\sin^2 \beta \times \cos \beta \times \sin 2\varphi_1}{(1 - \sin^2 \beta \times \sin^2 \varphi_1)^2}$$

Both axes A of the auxiliary shaft joints
must be parallel

The more the angle of inclination β increases, the more the maximum acceleration α and the accelerating moment M_α become; therefore, in practice, $\beta \leq 45$ deg.

Section 6

ELECTRICAL/ELECTRONIC

Symbol Definitions

These symbols are generally accepted as standard to represent electrical quantities.

Symbol	Definition
C	capacitance, farads
E	electromotive force, volts
E_{eff} or E_{rms}	effective or rms voltage
E_{max}	peak voltage
f	frequency, hertz
fr	resonant frequency, hertz
G	conductance, siemens
I	current, amperes
i	instantaneous current, amperes
I_{eff} or I_{rms}	effective or rms current
I_{max}	peak current
L	inductance, henries
λ	wavelength
M	mutual inductance, henries
O	instantaneous voltage
P	power, watts
PF	power factor
Q	figure of merit (quality)
R	resistance, ohms
θ	phase angle, degrees
T	time, seconds
X	reactance, ohms
X_c	capacitive reactance, ohms
X_L	inductive reactance, ohms
VA	apparent power, volt-amperes
Z	impedance, ohms

Ohm's Law

For AC Circuits

Known values	Formulas for determining unknown values of			
	I	Z	E	P
$I\&Z$			IZ	$I^2 Z \cos\theta$
$I\&E$		$\dfrac{E}{I}$		$IE\cos\theta$
$I\&P$		$\dfrac{P}{I^2\cos\theta}$	$\dfrac{P}{I\cos\theta}$	
$Z\&E$	$\dfrac{E}{Z}$			$\dfrac{E^2\cos\theta}{Z}$
$Z\&P$	$\sqrt{\dfrac{P}{Z\cos\theta}}$		$\sqrt{\dfrac{PZ}{\cos\theta}}$	
$E\&P$	$\dfrac{P}{E\cos\theta}$	$\dfrac{E^2\cos\theta}{P}$		

For DC Circuits

Known values	Formulas for determining unknown values of			
	I	R	E	P
$I\&R$			IR	$I^2 R$
$I\&E$		$\dfrac{E}{I}$		EI
$I\&P$		$\dfrac{P}{I^2}$	$\dfrac{P}{I}$	
$R\&E$	$\dfrac{E}{R}$			$\dfrac{E^2}{R}$
$R\&P$	$\sqrt{\dfrac{P}{R}}$		$\sqrt{PR}$	
$E\&P$	$\dfrac{P}{E}$	$\dfrac{E^2}{P}$		

Wire Chart

Current Capacity and Length vs Voltage Drop

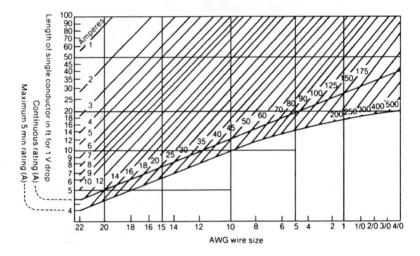

Ref.: *Specialties Handbook, 1964.* Copyright © 1964, Avionics Specialties, Inc., Charlottesville, VA. Reproduced with permission of Avionics Specialties.

Resistor Color Codes

Color Code for Small Resistors—Military and EIA

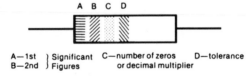

A—1st ⎫ Significant   C—number of zeros   D—tolerance
B—2nd ⎭ Figures   or decimal multiplier

Color	Significant figure	Multiplying value
Black	0	1
Brown	1	10
Red	2	100
Orange	3	1,000
Yellow	4	10,000
Green	5	100,000
Blue	6	1,000,000
Violet	7	10,000,000
Gray	8	100,000,000
White	9	1,000,000,000
Gold	±5% tolerance	0.1
Silver	±10% tolerance	0.01
No color	±20% tolerance	——

Resistor Color Codes, continued

Resistors available

Decimal multiples of						
5%				10%		20%
1.0	1.8	3.3	5.6	1.0	3.3	1.0
1.1	2.0	3.6	6.2	1.2	3.9	1.5
1.2	2.2	3.9	6.3	1.5	4.7	2.2
1.3	2.4	4.3	7.5	1.8	5.6	3.3
1.5	2.7	4.7	8.2	2.2	6.8	4.7
1.6	3.0	5.1	9.1	2.7	8.2	6.8

Properties of Insulating Materials

Insulating material	Dielectric constant (60 Hz)	Dielectric strength, V/mil	Resistivity, ohm-cm
Air normal pressure	1	19.8–22.8	
Amber	2.7–2.9	2300	Very high
Asphalts	2.7–3.1	25–30	——
Casein-moulded	6.4	400–700	Poor
Cellulose-acetate	6–8	250–1000	4.5×10^{10}
Ceresin wax	2.5–2.6	——	
Fibre	2.5–5	150–180	5×10^9
Glass-electrical	4–5	2000	8×10^{14}
Hallowax	3.4–3.8	——	$10^{13} \times 10^{14}$
Magnesium silicate	5.9–6.4	200–240	$>10^{14}$
Methacrylic resin	2.8	——	——
Mica	2.5–8	——	2×10^{17}
Micalex 364	6–8	350	——
Nylon	3.6	305	10^{13}
Paper	2–2.6	1250	——
Paraffin oil	2.2	381	
Paraffin wax	2.25	203–305	10^{16}
Phenol-yellow	5.3	500	——
Phenol-black moulded	5.5	400–500	——
Phenol-paper base	5.5	650–750	$10^{10} \times 10^{13}$
Polyethylene	2.25	1000	10^{17}
Polystyrene	2.5	508–706	10^{17}
Polyvinyl chloride	2.9–3.2	400	10^{14}
Porcelain—wet process	6.5–7	150	——
Porcelain—dry process	6.2–7.5	40–100	5×10^8
Quartz—fused	3.5–4.2	200	$10^{14}, 10^{18}$

(continued)

Properties of Insulating Materials, continued

Insulating material	Dielectric constant (60 Hz)	Dielectric strength, V/mil	Resistivity, ohm-cm
Rubber—hard	2–3.5	450	10^{12}, 10^{15}
Shellac	2.5–4	900	10^{16}
Steatite—commercial	4.9–6.5	——	——
Steatite—low-loss	4.4	150–315	10^{14}, 10^{15}
Titanium dioxide	90–170	100–210	10^{13}, 10^{14}
Varnished cloth	2–2.5	450–550	——
Vinyl resins	4	400–500	10^{14}
Wood—dry oak	2.5–6.8	——	——

Connectors

Frequently Used Connectors

Type	Governing spec	Coupling	Max freq.,[a] GHz	Voltage rating[b]	Relative cost	Overall size
SMC	MIL-C-39012	Thread	10	500	Low	Micro
SMB	MIL-C-39012	Snap-on	4	500	Low	Micro
SMA	MIL-C-39012	Thread	12.4	500	Med	Submin
TPS	MIL-C-55235	Bayonet	10	500	Med	Submin
TNC	MIL-C-39012	Thread	11	500	Med	Min
BNC	MIL-C-39012	Bayonet	4	500	Low	Min
N	MIL-C-39012	Thread	11	1000	Med-Low	Medium
SC	MIL-C-39012	Thread	11	1500	Med	Medium
C	MIL-C-39012	Bayonet	11	1500	Med	Medium
QDS	MIL-C-18867	Snap-on	11	1500	Med	Medium
HN	MIL-C-3643	Thread	2.5	5000	Med	Medium
LT(LC)	MIL-C-26637	Thread	4	5000	High	Large
QL	MIL-C-39012	Thread	5	5000	High	Large

[a] Maximum recommended operating frequency.
[b] Volts rms at sea level, tested at 60 Hz and 5 MHz (derate by a factor of 4 at 70,000-ft altitude).

High-Precision Connectors

Size	Governing spec	Coupling	Max freq.,[a] GHz
2.75 mm	NIST	Thread	40.0
3.5 mm	NIST	Thread	26.5
7 mm	IEEE 287	Thread	18
14 mm	IEEE 287	Thread	8.5

[a] Maximum recommended operating frequency.

Interface Connector Pin Assignments per RS-232-C

Pin number	Circuit	Description
1	AA	Protective ground
2	BA	Transmitted data
3	BB	Received data
4	CA	Request to send
5	CB	Clear to send
6	CC	Data set ready
7	AB	Signal ground (common return)
8	CF	Received line signal detector
9	——	(Reserved for data set testing)
10	——	(Reserved for data set testing)
11	——	Unassigned
12	SCF	Sec. rec'd. line sig. detector
13	SCB	Sec. clear to send
14	SBA	Secondary transmitted data
15	DB	Transmission signal element timing (DCE source)
16	SBB	Secondary received data
17	DD	Receiver signal element timing (DCE source)
18	——	Unassigned
19	SCA	Secondary request to send
20	CD	Data terminal ready
21	CG	Signal quality detector
22	CE	Ring indicator
23	CH/CI	Data signal rate selector (DTE/DCE source)
24	DA	Transmit signal element timing (DTE source)
25	——	Unassigned

Resistor, Capacitor, Inductance Combinations
Parallel Combinations of Resistors, Capacitors, and Inductors

Parallel combination	Impedance, ohms $(Z = R + jX)$	Magnitude of impedance, ohms $(|Z| = \sqrt{R^2 + X^2})$
R_1, R_2	$\dfrac{R_1 R_2}{R_1 + R_2}$	$\dfrac{R_1 R_2}{R_1 + R_2}$
C_1, C_2	$-j\dfrac{1}{\omega(C_1 + C_2)}$	$\dfrac{1}{\omega(C_1 + C_2)}$
L, R	$\dfrac{\omega^2 L^2 R + j\omega L R^2}{\omega^2 L^2 + R^2}$	$\dfrac{\omega L R}{\sqrt{\omega^2 L^2 + R^2}}$
R, C	$\dfrac{R - j\omega R^2 C}{1 + \omega^2 R^2 C^2}$	$\dfrac{R}{\sqrt{1 + \omega^2 R^2 C^2}}$
L, C	$+j - \dfrac{\omega L}{1 - \omega^2 LC}$	$\dfrac{\omega L}{1 - \omega^2 LC}$
$L_1(M)L_2$	$+j\omega\dfrac{L_1 L_2 - M^2}{L_1 + L_2 \mp 2M}$	$\omega\dfrac{L_1 L_2 - M^2}{L_1 + L_2 \mp 2M}$
L, C, R	$\dfrac{1}{R} - j\left(\omega C - \dfrac{1}{\omega L}\right)$ $\div \left(\dfrac{1}{R}\right)^2 + \left(\omega C - \dfrac{1}{\omega L}\right)^2$	$R \Big/ \sqrt{1 + R^2\left(\omega C - \dfrac{1}{\omega L}\right)^2}$

Parallel combination	Phase angle, rad $[\phi = \tan^{-1}(X/R)]$	Admittance, siemens $(Y = 1/Z)$
R_1, R_2	0	$\dfrac{R_1 + R_2}{R_1 R_2}$
C_1, C_2	$-\dfrac{\pi}{2}$	$+j\omega(C_1 + C_2)$
L, R	$\tan^{-1}\dfrac{R}{\omega L}$	$\dfrac{1}{R} - \dfrac{j}{\omega L}$
R, C	$\tan^{-1}(-\omega RC)$	$\dfrac{1}{R} + j\omega C$
L, C	$\pm\dfrac{\pi}{2}$	$j\left(\omega C - \dfrac{1}{\omega L}\right)$
$L_1(M)L_2$	$\pm\dfrac{\pi}{2}$	$-j\dfrac{1}{\omega}\left(\dfrac{L_1 + L_2 \mp 2M}{L_1 L_2 - M^2}\right)$
L, C, R	$\tan^{-1} - R\left(\omega C - \dfrac{1}{\omega L}\right)$	$\dfrac{1}{R} + j\left(\omega C - \dfrac{1}{\omega L}\right)$

Resistor, Capacitor, Inductance Combinations, continued

Series Combinations of Resistors, Capacitors, and Inductors

Series combination	Impedance, ohms $(Z = R + jX)$	Magnitude of impedance, ohms $(\lvert Z \rvert = \sqrt{R^2 + X^2})$
R	R	R
L	$+j\omega L$	ωL
C	$-j(1/\omega C)$	$1/\omega C$
$R_1 + R_2$	$R_1 + R_2$	$R_1 + R_2$
$L_1(M)L_2$	$+j\omega(L_1 + L_2 \pm 2M)$	$\omega(L_1 + L_2 \pm 2M)$
$C_1 + C_2$	$-j\dfrac{1}{\omega}\left(\dfrac{C_1 + C_2}{C_1 C_2}\right)$	$\dfrac{1}{\omega}\left(\dfrac{C_1 + C_2}{C_1 C_2}\right)$
$R + L$	$R + j\omega L$	$\sqrt{R^2 + \omega^2 L^2}$
$R + C$	$R - j\dfrac{1}{\omega C}$	$\sqrt{\dfrac{\omega^2 C^2 R^2 + 1}{\omega^2 C^2}}$
$L + C$	$+j\left(\omega L - \dfrac{1}{\omega C}\right)$	$\left(\omega L - \dfrac{1}{\omega C}\right)$
$R + L + C$	$R + j\left(\omega L - \dfrac{1}{\omega C}\right)$	$\sqrt{R^2 + \left(\omega L - \dfrac{1}{\omega C}\right)^2}$

Series combination	Phase angle, rad $[\phi = \tan^{-1}(X/R)]$	Admittance, siemens $(Y = 1/Z)$
R	0	$1/R$
L	$+\pi/2$	$-j(1/\omega L)$
C	$-\pi/2$	$j\omega C$
$R_1 + R_2$	0	$1/(R_1 + R_2)$
$L_1(M)L_2$	$+\pi/2$	$-j/\omega(L_1 + L_2 \pm 2M)$
$C_1 + C_2$	$-\dfrac{\pi}{2}$	$j\omega\left(\dfrac{C_1 C_2}{C_1 + C_2}\right)$
$R + L$	$\tan^{-1}\dfrac{\omega L}{R}$	$\dfrac{R - j\omega L}{R^2 + \omega^2 L^2}$
$R + C$	$-\tan^{-1}\dfrac{1}{\omega RC}$	$\dfrac{\omega^2 C^2 R + j\omega C}{\omega^2 C^2 R^2 + 1}$
$L + C$	$\pm\dfrac{\pi}{2}$	$-\dfrac{j\omega C}{\omega^2 LC - 1}$
$R + L + C$	$\tan^{-1}\left(\dfrac{\omega L - 1/\omega C}{R}\right)$	$\dfrac{R - j(\omega L - 1/\omega C)}{R^2 + (\omega L - 1/\omega C)^2}$

Dynamic Elements and Networks

Element or System

$$G(s)$$

Integrating Circuit

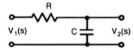

$$\frac{V_2(s)}{V_1(s)} = \frac{1}{RCS + 1}$$

Differentiating Circuit

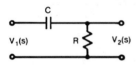

$$\frac{V_2(s)}{V_1(s)} = \frac{RCS}{RCS + 1}$$

Differentiating Circuit

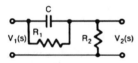

$$\frac{V_2(s)}{V_1(s)} = \frac{s + 1/R_1 C}{s + (R_1 + R_2)/R_1 R_2 C}$$

Lead-Lag Filter Circuit

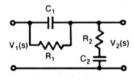

$$\frac{V_2(s)}{V_1(s)} = \frac{(1 + s\tau_a)(1 + s\tau_b)}{\tau_a \tau_b s^2 + (\tau_a + \tau_b + \tau_{ab})s + 1}$$

$$= \frac{(1 + s\tau_a)(1 + s\tau_b)}{(1 + s\tau_1)(1 + s\tau_2)}$$

$$\tau_a = R_1 C_1$$

$$\tau_b = R_2 C_2$$

$$\tau_{ab} = R_1 C_1$$

$$\tau_1 \tau_2 = \tau_a \tau_b$$

$$\tau_1 + \tau_2 = \tau_a + \tau_b + \tau_{ab}$$

Dynamic Elements and Networks, continued

DC-Motor, Field Controlled

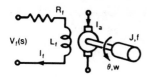

$$\frac{\theta(s)}{V_f(s)} = \frac{K_m}{s(Js + f)(L_f s + R_f)}$$

DC-Motor, Armature Controlled

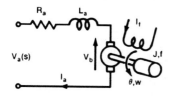

$$\frac{\theta(s)}{V_a(s)} = \frac{K_m}{s[(R_a + L_a s)(Js + f) + K_b K_m]}$$

Laplace Transforms

The Laplace transform of a function $f(t)$ is defined by the expression

$$F(p) = \int_0^\infty f(t)e^{-pt}\,dt$$

If this integral converges for some $p = p_0$, real or complex, then it will converge for all p such that $\mathrm{Re}\,p > \mathrm{Re}\,p_0$.

The inverse transform may be found by

$$f(t) = (j2\pi)^{-1} \int_{c-j\infty}^{c+j\infty} F(z)e^{tz}\,dz \qquad t > 0$$

where there are no singularities to the right of the path of integration.

Laplace Transforms, continued

General Equations

	Function	Transform[a]
Shifting theorem	$f(t - a), \; f(t) = 0, \; t < 0$	$e^{-ap}F(p), \; a > 0$
Convolution	$\int_0^t f_1(\lambda)f_2(t - \lambda)\,d\lambda$	$F_1(p)F_2(p)$
Linearity	$a_1 f_1(t) + a_2 f_2(t),$ $(a_1, a_2 \text{ const})$	$a_1 F_1(p) + a_2 F_2(p)$
Derivative	$df(t)/dt$	$-f(0) + pF(p)$
Integral	$\int f(t)\,dt$	$p^{-1}\left[\int f(t)\,dt\right]_{t=0} + [f(p)/p],$ $\mathrm{Re}\,p > 0$
Periodic function	$f(t) = f(t + r)$	$\int_0^r f(\lambda)e^{-p\lambda}\,d\lambda/(1 - e^{-pr}), \; r > 0$
	$f(t) = -f(t + r)$	$\int_0^r f(\lambda)e^{-\lambda}\,d\lambda/(1 + e^{-pr}), \; r > 0$
	$f(at), \; a > 0$	$F(p/a)/a$
	$e^{at}f(t)$	$F(p - a), \; \mathrm{Re}\,p > \mathrm{Re}\,a$
	$t^n f(t)$	$(-1)^n[d^n F(p)/dp^n]$
Final-value theorem	$f(\infty)$	$\lim_{p \to 0} pF(p)$
Initial-value theorem	$f(0+)$	$\lim_{p \to \infty} pF(p)$

[a]$F(p)$ denotes the Laplace transform of $f(t)$.

Miscellaneous Functions[a]

	Function	Transform
Step	$u(t - a) = 0, \; 0 \le t < a$ $= 1, \; t \ge a$	e^{-ap}/p
Impulse	$\delta(t)$	1
	$t^a, \; \mathrm{Re}\,a > -1$	$\Gamma(a + 1)/p^{a+1}$
	e^{at}	$1/(p - a), \; \mathrm{Re}\,p > \mathrm{Re}\,a$
	$t^a e^{bt}, \; \mathrm{Re}\,a > -1$	$\Gamma(a + 1)/(p - b)^{a+1}, \; \mathrm{Re}\,p > \mathrm{Re}\,b$
	$\cos at$	$p/(p^2 + a^2)$ $\left.\right\} \mathrm{Re}\,p > |\mathrm{Im}\,a|$
	$\sin at$	$a/(p^2 + a^2)$

(continued)

[a]For an extensive listing, refer to A. Erdéyli, ed., *Tables of Integral Transforms*, Vol. 1, Bateman Manuscript Project, New York: McGraw–Hill Book Co., 1954.

Laplace Transforms, continued

Miscellaneous Functions,[a] continued

Function	Transform
$\cosh at$ $\sinh at$	$\left.\begin{array}{l} p/(p^2 - a^2) \\ 1/(p^2 - a^2) \end{array}\right\} \mathrm{Re}\,p > |\mathrm{Re}\,a|$
$\ell_n t$	$-(\gamma + \ell_n p)/p$, γ is Euler's constant $= 0.57722$
$1/(t + a)$, $a > 0$	$e^{ap} E_1(ap)$
e^{-at2}	$\frac{1}{2}(\pi/a)^{1/2} e^{p^2/4a} \mathrm{erfc}[p/2(a)^{1/2}]$
Bessel function $J_v(at)$, $\mathrm{Re}\,v > -1$	$r^{-1}[(r - p)/a]^v$, $r = (p^2 + a^2)^{1/2}$, $\mathrm{Re}\,p > |\mathrm{Re}\,a|$
Bessel function $I_v(at)$, $\mathrm{Re}\,v > -1$	$R^{-1}[(R - p)/a]^v$, $R = (p^2 - a^2)^{1/2}$, $\mathrm{Re}\,p > |\mathrm{Re}\,a|$

[a]For an extensive listing, refer to A. Erdéyli, ed., *Tables of Integral Transforms*, Vol. 1, Bateman Manuscript Project, New York: McGraw–Hill Book Co., 1954.

Inverse Transforms[a]

Transform	Function
1	$\delta(t)$
$1/(p + a)$	e^{-at}
$1/(p + a)^v$, $\mathrm{Re}\,v > 0$	$t^{v-1} e^{-at}/\Gamma(v)$
$1/[(p + a)(p + b)]$	$(e^{-at} - e^{-bt})/(b - a)$
$p/[(p + a)(p + b)]$	$(ae^{-at} - be^{-bt})/(a - b)$
$1/(p^2 + a^2)$	$a^{-1} \sin at$
$1/(p^2 - a^2)$	$a^{-1} \sinh at$
$p/(p^2 + a^2)$	$\cos at$
$p/(p^2 - a^2)$	$\cosh at$
$1/(p^2 + a^2)^{1/2}$	$J_o(at)$
e^{-ap}/p	$u(t - a)$
e^{-ap}/p^v, $\mathrm{Re}\,v > 0$	$(t - a)^{v-1} u(t - a)/\Gamma(v)$
$(1/p)e^{-a/p}$	$J_0[2(at)^{1/2}]$
$\left.\begin{array}{l}(1/p^v)e^{-a/p} \\ (1/p^v)e^{a/p}\end{array}\right\} \mathrm{Re}\,v > 0$	$(t/a)^{(v-1)/2} J_{v-1}[2(at)^{1/2}]$ $(t/a)^{(v-1)/2} I_{v-1}[2(at)^{1/2}]$
$(1/p)\,\ell_n p$	$-\gamma - \ell_n t$, $\gamma = 0.57722$

[a]Refer to A. Erdéyli, ed., *Tables of Integral Transforms*, Vol. 1, Bateman Manuscript Project, New York: McGraw–Hill Book Co., 1954.

Voice Transmission Facilities

Number of voice circuits, present and future

Transmission method	Maximum frequency used	Voice circuits per system	Maximum voice circuits per route
Carrier on paired cable	260 kHz	12–24	1,000–2,000
Digital on paired cable	1.5 MHz	24	4,800
Carrier on coaxial cable	2.7 MHz	600	1,800
	8.3 MHz	1860	5,580
	18.0 MHz	3600	32,900
Carrier on microwave radio	4.2 GHz	600	3,000
	6.4 GHz	1800	14,000
	11.7 GHz	100	400
Satellite	30 GHz	1500	20,000
Millimeter wave guide	100 GHz	5000	250,000
Optical-guide laser	1,000,000 GHz	?	10,000,000

Ref.: "The Changing Criteria for Engineering Designs" by P. S. Meyers, SAE-690303. Copyright © 1969, the Society of Automotive Engineers, Warrendale, PA. Reproduced with permission of the Society of Automotive Engineers.

Electromagnetic Spectrum

Electromagnetic Wave Spectrum Wave Length vs Frequency

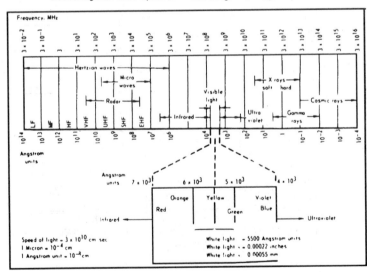

$$dB = 10 \log_{10} \frac{P_1}{P_2} \qquad dB = 20 \log_{10} \frac{V_1}{V_2} = 20 \log_{10} \frac{I_1}{I_2}$$

Ref.: *Weight Engineers Handbook*, Revised 1976. Copyright © 1976, the Society of Allied Weight Engineers, La Mesa, CA. Reproduced with permission of the Society of Allied Weight Engineers.

Antennas

Patterns, Gains, and Areas of Typical Antennas

Type	Configuration	Pattern	Power gain over isotropic	Effective area
Electric doublet		$\cos\theta$	1.5	$1.5\lambda^2/4\pi$
Magnetic doublet or loop		$\sin\theta$	1.5	$1.5\lambda^2/4\pi$
Half-wave dipole		$\dfrac{\cos(\pi/2\,\sin\theta)}{\cos\theta}$	1.64	$1.64\lambda^2/4\pi$
Half-wave dipole and screen		$2\sin(S°\cos\beta)$	6.5	$1.64\lambda^2/4\pi$
Turnstile array	$n=5$	$\dfrac{\sin(nS°/2\,\sin\beta)}{n\sin(S°/2\,\sin\beta)}$	n or $2L/\lambda$	$n\lambda^2/4\pi$ or $L\lambda/2\pi$
Loop array	$n=5$	$\dfrac{\cos\beta\,\sin(nS°/2\,\sin\beta)}{n\sin(S°/2\,\sin\beta)}$	n or $2L/\lambda$	$n\lambda^2/4\pi$ or $L\lambda/2\pi$
Optimum horn $L \geq a^2/\lambda$		Half-power width $70\,\lambda/a$ deg (H plane) $51\,\lambda/b$ deg (E plane)	$10\,ab/\lambda^2$	$0.81ab$
Parabola		Half-power width $70\,\lambda/d$ deg	$2\pi d^2/\lambda^2$	$d^2/2$

Ref.: *NAB Engineering Handbook*, 5th Edition. Copyright © 1960, McGraw–Hill, New York. Reproduced with permission of The McGraw–Hill Companies.

Frequency Classifications

ELF	30–300 Hz	Extremely low frequency
VF	300 Hz–3 kHz	Voice frequency
VLF	3–30 kHz	Very low frequency
LF	30–300 kHz	Low frequency
MF	300 kHz–3.00 MHz	Medium frequency
HF	3.00–30 MHz	High frequency
VHF	30–300 MHz	Very high frequency
UHF	300 MHz–3 GHz	Ultrahigh frequency
SHF	3–30 GHz	Superhigh frequency
EHF	30–300 GHz	Extremely high frequency

Radar Letter-Band Nomenclature

Band designation	Nominal frequency range	Specific radio location (radar) bands based on ITU assignments for region 2
HF	3–30 MHz	
VHF	30–300 MHz	138–144 MHz
		216–225
UHF	300–1000 MHz	420–450 MHz
		890–942
L	1–2 GHz	1,215–1,400 MHz
S	2–4 GHz	2,300–2,500 MHz
		2,700–3,700
C	4–8 GHz	5,250–5,925 MHz
X	8–12 GHz	8,500–10,680 MHz
K_u	12–18 GHz	13.4–14.0 GHz
		15.7–17.7
K	18–27 GHz	24.05–24.25 GHz
K_a	27–40 GHz	33.4–36.0 GHz
V	40–75 GHz	59–64 GHz
W	75–110 GHz	76–81 GHz
		92–100
mm	110–300 GHz	126–142 GHz
		144–149
		231–235
		238–248

Radar Range Equation

Maximum radar range can be calculated from the following simplified equation:

$$R_{max} = K \sqrt{\frac{P_T \cdot \lambda^2 \cdot G_0^2 \cdot \gamma \cdot F^{1/3} \cdot A_E}{NF}}$$

where

R_{max} = range, n mile
P_T = peak transmitter power, W
λ = wavelength, cm
G_0 = antenna gain, dB
γ = pulse duration, μs
F = pulse repetition rate, cycles per s
A_E = target size, m^2
NF = receiver noise figure, dB
K = 5.5

For detailed information, see *Radar Cross Section Lectures* by Allen E. Fuhs, published by AIAA.

U.S. Electronic Equipment Designations

All U.S. military electronic equipment are assigned an identifying alphanumeric designation that can be used to determine the platform the equipment was designed to be operated on, the type of equipment, and its function. This system is commonly called the AN designation system, although its formal name is the Joint Electronics Type Designation System (JETDS). The letters AN preceding the equipment indicators formerly meant Army/Navy but are now considered to be an exclusive letter set that can only be used to indicate formally designated DOD equipment.

Letters following the AN designation numbers provide added information about equipment. Suffixes A, B, C, etc., indicate the latest modification. AN/ALR-67B would indicate the second modification to the system. The letter "V" indicates variable configurations are available. The letter "X" indicates developmental status. A double parenthesis ( ) indicates a generic system that has not yet received a formal designation, i.e., AN/ALQ-( ), or AN/ALQ-165( ).

Example: AN/ALQ-165(V)

AN/ = A formally designated military system (DOD designation)
A = Mounted on a piloted aircraft (platform installation)
L = A countermeasures set (equipment type)
Q = Special or combination of purposes (equipment function or purpose)
165 = 165th model of this type designated (model number)
(V) = Variable configurations (suffix)

U.S. Electronic Equipment Designations, continued

Equipment indicators

A. Platform installation

A	Piloted aircraft	S	Water
B	Underwater mobile, submarine	T	Ground transportable
		U	General utility
D	Pilotless carrier	V	Vehicular (ground)
F	Fixed ground	W	Water surface and underwater combination
G	General ground use		
K	Amphibious	Z	Piloted-pilotless airborne vehicle combination
M	Mobile (ground)		
P	Portable		

B. Equipment type

A	Invisible light, heat radiation	N	Sound in air
C	Carrier	P	Radar
D	Radiac	Q	Sonar and underwater sound
G	Telegraph or teletype		
I	Interphone and public address	R	Radio
		S	Special or combinations of types
J	Electromechanical or inertial wire covered	T	Telephone (wire)
K	Telemetering	V	Visual and visible light
L	Countermeasures	W	Armament
M	Meteorological	X	Facsimile or television
		Y	Data processing

C. Equipment function or purpose

B	Bombing	N	Navigational aids
C	Communications	Q	Special or combination of purposes
D	Direction finder, reconnaissance and/or surveillance		
		R	Receiving, passive detecting
E	Ejection and/or release	S	Detecting and/or range and bearing, search
G	Fire control or search-light directing		
		T	Transmitting
H	Recording and/or reproducing	W	Automatic flight or remote control
K	Computing	X	Identification and recognition
M	Maintenance and/or test assemblies		
		Y	Surveillance and control

Ref.: *The International Countermeasures Handbook*, 8th Edition, published by EW Communications, Inc., Palo Alto, CA, 1982.

Section 7

AIRCRAFT DESIGN

Vehicle Definitions

Geometry

The following figures and formulas provide an introduction on geometric relationships concerning vehicle physical dimensions. These dimensions are used throughout this section.

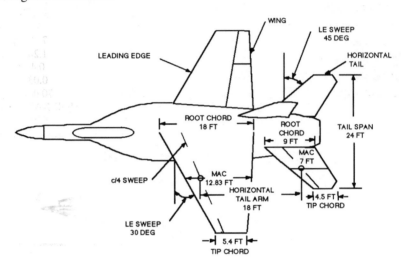

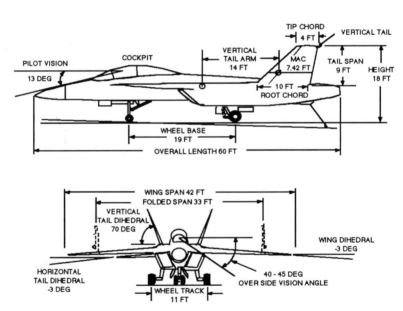

Vehicle Definitions, continued

Geometry	Unit	Wing	H-Tail	V-Tail
LE sweep angle	deg	30.0	45.0	45.0
c/4 sweep angle	deg	23.1	42.2	33.6
Reference area	ft^2	491.4	162	63 each
Projected span	ft	42	24	9
m.a.c.	ft	12.83	7	7.42
Aspect ratio	AR	3.59	3.55	1.28
Taper ratio	λ	0.3	0.5	0.4
Thickness ratio	t/c	0.05	0.04	0.03
Dihedral	Γ	−3 deg	−3 deg	70 deg
Airfoil	——	MOD NACA 65A	MOD NACA 65A	MOD NACA 65A
Tail volume	$\bar{V}$	n/a	0.462	0.28

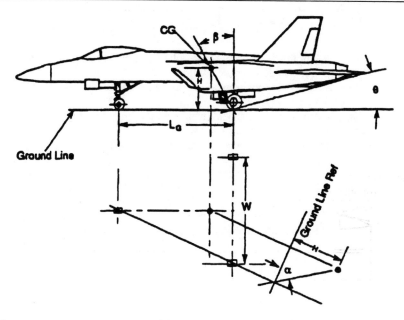

where

L_G = wheel base
W = wheel track
H = height c.g. to ground reference
β = tip back angle
θ = tail down angle
α = turn over angle

Gear in normal static position
Use most aft c.g. for β, keep $\beta > \theta$
Use most forward c.g. for α
Use landing weight c.g. for θ

Vehicle Definitions, continued

The following definitions and equations apply to trapezoidal planforms, as illustrated here.

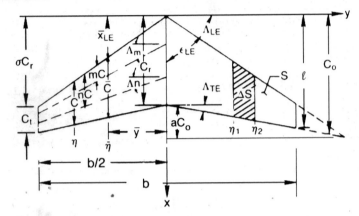

C_o = overall length of zero-taper-ratio planform having same leading- and trailing-edge sweep as subject planform

σ = ratio of chordwise position of leading edge at tip to the root chord length
 = $(b/2)\tan \Lambda_{LE}(1/C_r)$

η_1, η_2 = span stations of boundary of arbitrary increment of wing area

Λ_m, Λ_n = sweep angles of arbitrary chordwise locations

m, n = nondimensional chordwise stations in terms of C

General

$$\eta = \frac{y}{b/2}$$

$$\lambda = C_t/C_r$$

$$C = C_r[1 - \eta(1 - \lambda)]$$

$$\tan \Lambda = 1/\tan \epsilon$$

$$x_{LE} = (b/2)\eta \tan \Lambda_{LE}$$

$$\ell = \frac{b}{2} \tan \Lambda_{LE} + C_t = C_r \frac{1 - a\lambda}{1 - a}$$

Area

$$S = \frac{b^2}{AR} = \frac{b}{2}C_r(1 + \lambda) = \frac{b}{2}C_o(1 - a)(1 - \lambda)$$

$$= \frac{C_o^2(1 - a)(1 - \lambda^2)}{\tan \Lambda_{LE}}$$

$$\Delta S = \frac{b}{2}C_r[2 - (1 - \lambda)(\eta_1 - \eta_2)]\frac{\eta_2 - \eta_1}{2}$$

Vehicle Definitions, continued

Aspect Ratio

$$AR = \frac{b^2}{S} = \frac{2b}{C_r(1+\lambda)} = \frac{4(1-\lambda)}{(1-a)(1+\lambda)\tan \Lambda_{LE}}$$

Cutout Factor

$$a = \frac{\tan \Lambda_{TE}}{\tan \Lambda_{LE}} = 1 - \frac{C_r(1-\lambda)}{(b/2)\tan \Lambda_{LE}} = 1 - \frac{4(1-\lambda)}{AR(1+\lambda)\tan \Lambda_{LE}}$$

Sweep Angles

$$\tan \Lambda_{LE} = \frac{1}{a}\tan \Lambda_{TE} = \frac{C_r(1-\lambda)}{(b/2)(1-a)} = \frac{4(1-\lambda)}{AR(1+\lambda)(1-a)}$$

$$= \frac{C_o(1-\lambda)}{b/2} = \frac{AR(1+\lambda)\tan \Lambda_{c/4} + (1-\lambda)}{AR(1+\lambda)} = \frac{4\tan \Lambda_{c/4}}{3+a}$$

$$\tan \Lambda_m = \tan \Lambda_{LE}[1 - (1-a)m]$$

$$\tan \Lambda_m = \tan \Lambda_{LE} - \frac{4m}{AR}\left(\frac{1-\lambda}{1+\lambda}\right)$$

$$\cos \Lambda_m = (\tan \Lambda_{LE})^{-1}\left\{\left(\frac{1}{\tan \Lambda_{LE}}\right)^2 + [1 - (1-a)m]^2\right\}^{-\frac{1}{2}}$$

Mean Aerodynamic Chord (m.a.c.)

$$\bar{C} = \frac{2}{S}\int_o^{b/2} C^2\,dy = \frac{2}{3}C_r\left(1 + \frac{\lambda^2}{1+\lambda}\right)$$

$$= \frac{2}{3}C_o(1-a)\left(1 + \frac{\lambda^2}{1+\lambda}\right)$$

$$= \frac{4}{3}\left(\frac{S}{AR}\right)^{\frac{1}{2}}\left[1 - \frac{\lambda}{(1+\lambda)^2}\right]$$

$$= \frac{2}{3}\left(C_r + C_t - \frac{C_r \times C_t}{C_r + C_t}\right)$$

$$\eta = \frac{2}{S}\int_o^{b/2} Cy\,dy = \frac{1 - (\bar{C}/C_r)}{1-\lambda} = \frac{1}{3}\left(\frac{1+2\lambda}{1+\lambda}\right)$$

$$\bar{x}_{LE} = \bar{y}\tan \Lambda_{LE}$$

Vehicle Definitions, continued

Root Chord

$$C_r = \frac{S}{(b/2)(1 + \lambda)} = \frac{4(b/2)}{AR(1 + \lambda)}$$

Chordwise Location of Leading Edge at Tip

$$\sigma = \frac{AR}{4}(1 + \lambda)\tan \Lambda_{LE}$$

Force-Velocity

Airplane Axis System

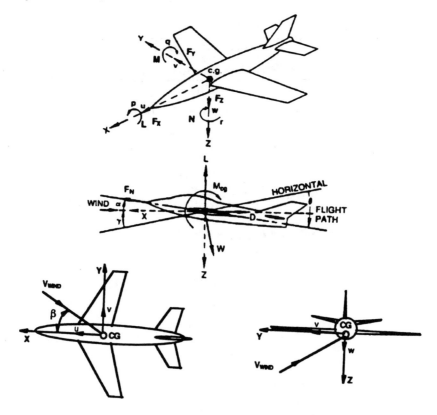

Axis	Force along	Moment about	Linear velocity	Ang. disp.	Ang. vel.	Inertia	Ang. of attack
X	F_x	L	u	ϕ	p	I_x	σ
Y	F_y	M	v	θ	q	I_y	α
Z	F_z	N	w	ψ	r	I_z	β

Aerodynamics

Basic Aerodynamic Relationships

AR = aspect ratio = b^2/S

C_D = drag coefficient = $D/qS = C_{DO} + C_{Di}$

C_{Di} = induced drag coefficient = $C_L^2/(\pi AR e)$

C_L = lift coefficient = L/qS

C_ℓ = rolling-moment coefficient = rolling moment/qbS

C_m = pitching-moment coefficient = pitching moment/qcS

C_n = yawing-moment coefficient = yawing moment/qbS

C_y = side-force coefficient = side force/qS

D = drag = $C_D qS$

d = equivalent body diameter = $\sqrt{4A_{MAX}/\pi}$

$\overline{FR}$ = fineness ratio = ℓ/d

L = lift = $C_L qS$

M = Mach number = V/a

P = planform shape parameter = $S/b\ell$

q = dynamic pressure = $\frac{1}{2}(\rho V^2) = \frac{1}{2}(\rho a^2 M^2)$

R_n = Reynolds number = $V\ell\rho/\mu$

R_o = $d/2$ = equivalent body radius

$(t/c)_{RMS}$ = root-mean-square thickness ratio

$$(t/c)_{RMS} = \left[\frac{1}{b/2 - r} \int_r^{b/2} (t/c)^2 \, dy \right]^{\frac{1}{2}}$$

V = true airspeed = $V_e/\sigma^{1/2}$

$\bar{X}$ = chordwise location from apex to $\bar{C}/Z$ (equivalent to chordwise location of centroid of area)

$$\bar{X} = \frac{2}{S} \int_o^{b/2} c\left(x + \frac{c}{2}\right) dy$$

$\bar{X}_{LE}$ = chordwise location of leading edge of m.a.c.

$$\bar{X}_{LE} = x - \frac{\bar{c}}{2}$$

$\bar{Y}$ = spanwise location of $\bar{C}$ (equivalent to spanwise location of centroid of area)

$$\bar{Y} = \frac{2}{S} \int_o^{b/2} cy \, dy$$

β = $\sqrt{M^2 - 1}$ (supersonic), $\sqrt{1 - M^2}$ (subsonic)

ϵ = complement to wing sweep angle = 90 deg$-\Lambda_{LE}$

η = nondimensional span station = $y/(b/2)$

λ = taper ratio, tip-to-root chord = C_t/C_r

σ = air density ratio = ρ/ρ_o

ν = kinematic viscosity = μ/ρ

Aerodynamics, continued

Symbols

a	= speed of sound
A_{MAX}	= maximum cross-sectional area
a.c.	= aerodynamic center
b	= wing span
C	= chord
$\bar{C}$	= mean aerodynamic chord (m.a.c.)
C_{DO}	= drag coefficient at zero lift
c.g.	= center of gravity
c.p.	= center of pressure
C_r	= root chord
C_t	= tip chord
d	= diameter
$(dA/dx)AFT$	= slope of aft end of configuration distribution curve
e	= Oswald (wing) efficiency factor
g	= acceleration due to gravity
i	= angle of incidence
K_{BODY}	= body wave-drag factor
K_{LE}	= wing shape factor
ℓ	= characteristic length
S, S_{REF}	= reference area
S_{EXP}	= exposed planform area
T	= temperature
t	= airfoil maximum thickness at span station y
t/c	= airfoil thickness ratio (parallel to axis of symmetry)
V_e	= equivalent velocity
x	= general chordwise location, parallel to plane of symmetry
y	= general spanwise location, perpendicular to plane of symmetry
α	= angle of attack, chord plane to relative wind
α_{Lo}	= angle of attack for zero lift
Γ	= dihedral angle
δ	= surface deflection angle
γ	= ratio of specific heats
Λ	= sweep-back angle
Λ_{LE}	= wing leading-edge sweep angle
Λ_{TE}	= wing trailing-edge sweep angle
Λ_{HL}	= flap-hinge-line sweep angle
μ	= coefficient of absolute viscosity
ρ	= density
R	= specific gas constant

Aerodynamics, continued

Speed of Sound vs Temperature

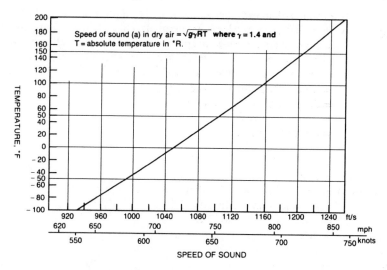

Dynamic Pressure (*q*) vs Mach Number

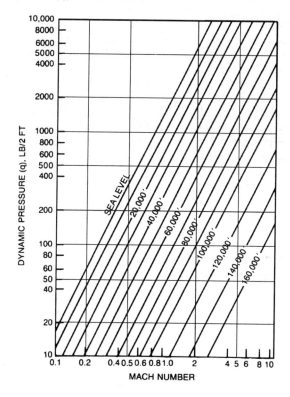

Aerodynamics, continued

Properties of the ICAO Standard Atmosphere

Standard atmosphere is a hypothetical vertical distribution of atmospheric temperature, pressure, and density which, by international or national agreement, is taken to be the representative of the atmosphere for the purpose of altimeter calculations, aircraft design, performance calculations, etc. The internationally accepted standard atmosphere is called the International Civil Aviation Organization (ICAO) Standard Atmosphere or the International Standard Atmosphere (ISA).

Standard atmosphere

h,	Temperature		p,	$\rho \cdot 10^4$	$\nu \cdot 10^4$	δ	σ			Knots	
ft	°R	K	in. Hg	slug/ft³	ft²s	(p/p_o)	(ρ/ρ_o)	$\sigma^{1/2}$	q/M^2	a	$a\sigma^{1/2}$
0	518.7	288.2	29.92	23.77	1.576	1.0000	1.0000	1.0000	1483	661.3	661.3
1,000	515.1	286.2	28.86	23.08	1.614	0.9644	0.9711	0.9854	1430	659.0	649.4
2,000	511.6	284.2	27.82	22.41	1.653	0.9298	0.9428	0.9710	1379	656.7	637.7
3,000	508.0	282.2	26.82	21.75	1.694	0.8962	0.9151	0.9566	1329	654.4	626.0
4,000	504.4	280.2	25.84	21.11	1.735	0.8637	0.8881	0.9424	1280	652.1	614.5
5,000	500.9	278.3	24.90	20.48	1.778	0.8320	0.8617	0.9282	1234	649.8	603.1
6,000	497.3	276.3	23.98	19.87	1.823	0.8014	0.8359	0.9142	1188	647.5	591.9
7,000	493.7	274.3	23.09	19.27	1.869	0.7716	0.8106	0.9003	1144	645.2	580.9
8,000	490.2	272.3	22.22	18.68	1.916	0.7428	0.7860	0.8866	1101	642.9	570.9
9,000	486.6	270.3	21.39	18.11	1.965	0.7148	0.7620	0.8729	1060	640.5	559.1
10,000	483.0	268.3	20.58	17.55	2.015	0.6877	0.7385	0.8593	1019	638.1	548.3
11,000	479.5	266.4	19.79	17.01	2.067	0.6614	0.7155	0.8459	980.5	635.8	537.8
12,000	475.9	264.4	19.03	16.48	2.121	0.6360	0.6932	0.8326	942.8	633.4	527.4
13,000	472.3	262.4	18.29	15.96	2.177	0.6113	0.6713	0.8193	906.3	631.1	517.1
14,000	468.8	260.4	17.58	15.45	2.234	0.5875	0.6500	0.8063	870.9	628.7	506.9
15,000	465.2	258.4	16.89	14.96	2.294	0.5643	0.6292	0.7933	836.6	626.3	496.8
16,000	461.6	256.4	16.22	14.47	2.355	0.5420	0.6090	0.7803	803.5	623.9	486.8
17,000	458.1	254.5	15.57	14.01	2.419	0.5203	0.5892	0.7676	771.3	621.4	477.0
18,000	454.5	252.5	14.94	13.55	2.485	0.4994	0.5699	0.7549	740.3	619.0	467.3
19,000	450.9	250.5	14.34	13.10	2.553	0.4791	0.5511	0.7424	710.2	616.6	457.8
20,000	447.4	248.6	13.75	12.66	2.624	0.4595	0.5328	0.7299	681.2	614.1	448.2
21,000	443.8	246.6	13.18	12.24	2.696	0.4406	0.5150	0.7176	653.1	611.7	439.0
22,000	440.2	244.6	12.64	11.83	2.772	0.4223	0.4976	0.7054	626.1	609.2	429.7
23,000	436.7	242.6	12.11	11.43	2.850	0.4046	0.4807	0.6933	599.9	606.8	420.7
24,000	433.1	240.6	11.60	11.03	2.932	0.3876	0.4642	0.6813	574.6	604.3	411.7

(continued)

Aerodynamics, continued

Standard atmosphere, continued

$h.$ ft	Temperature $^\circ$R	K	$p.$ in. Hg	$\rho . 10^4$ slug/ft^3	$\nu . 10^4$ ft^{2}s	δ (p/p_o)	σ (ρ/ρ_o)	$\sigma^{1/2}$	q/M^2	Knots a	$a\sigma^{1/2}$
25,000	429.5	238.6	11.10	10.65	3.016	0.3711	0.4481	0.6694	550.2	601.8	402.8
26,000	426.0	236.7	10.63	10.28	3.103	0.3552	0.4325	0.6576	526.6	599.3	394.1
27,000	422.4	234.7	10.17	9.918	3.194	0.3398	0.4173	0.6460	503.8	596.8	385.5
28,000	418.8	232.7	9.725	9.567	3.287	0.3250	0.4025	0.6344	481.8	594.2	377.0
29,000	415.3	230.7	9.297	9.225	3.385	0.3107	0.3881	0.6230	460.7	591.7	368.6
30,000	411.7	228.7	8.885	8.893	3.486	0.2970	0.3741	0.6117	440.2	589.2	360.4
31,000	408.1	226.7	8.488	8.570	3.591	0.2837	0.3605	0.6005	420.6	586.6	352.3
32,000	404.6	224.8	8.106	8.255	3.700	0.2709	0.3473	0.5893	401.6	584.0	344.2
33,000	401.0	222.8	7.737	7.950	3.813	0.2586	0.3345	0.5783	383.4	581.5	336.3
34,000	397.4	220.8	7.382	7.653	3.931	0.2467	0.3220	0.5674	365.8	578.9	328.5
35,000	393.9	218.8	7.041	7.365	4.053	0.2353	0.3099	0.5567	348.8	576.3	320.8
36,000	390.3	216.8	6.712	7.086	4.181	0.2243	0.2981	0.5460	332.6	573.6	313.2
36,089	390.0	216.7	6.683	7.061	4.192	0.2234	0.2971	0.5450	331.2	573.4	312.5
37,000	390.0	216.7	6.397	6.759	4.380	0.2138	0.2844	0.5332	317.0	573.4	305.7
38,000	390.0	216.7	6.097	6.442	4.596	0.2038	0.2710	0.5206	302.1	573.4	298.5
39,000	390.0	216.7	5.811	6.139	4.822	0.1942	0.2583	0.5082	287.9	573.4	291.4
40,000	390.0	216.7	5.538	5.851	5.059	0.1851	0.2462	0.4962	274.4	573.4	284.5
41,000	390.0	216.7	5.278	5.577	5.308	0.1764	0.2346	0.4844	261.5	573.4	277.8
42,000	390.0	216.7	5.030	5.315	5.570	0.1681	0.2236	0.4729	249.2	573.4	271.2
43,000	390.0	216.7	4.794	5.066	5.844	0.1602	0.2131	0.4616	237.5	573.4	264.7
44,000	390.0	216.7	4.569	4.828	6.132	0.1527	0.2031	0.4507	226.4	573.4	258.4
45,000	390.0	216.7	4.355	4.601	6.434	0.1455	0.1936	0.4400	215.8	573.4	252.3
46,000	390.0	216.7	4.151	4.385	6.750	0.1387	0.1845	0.4295	205.7	573.4	246.3
47,000	390.0	216.7	3.950	4.180	7.083	0.1322	0.1758	0.4193	196.0	573.4	240.4
48,000	390.0	216.7	3.770	3.983	7.432	0.1250	0.1676	0.4094	186.8	573.4	234.7
49,000	390.0	216.7	3.593	3.797	7.797	0.1201	0.1597	0.3996	178.0	573.4	229.1
50,000	390.0	216.7	3.425	3.618	8.181	0.1145	0.1522	0.3902	169.7	573.4	223.7
51,000	390.0	216.7	3.264	3.449	8.584	0.1091	0.1451	0.3809	161.7	573.4	218.4
52,000	390.0	216.7	3.111	3.287	9.007	0.1040	0.1383	0.3719	154.1	573.4	213.2
53,000	390.0	216.7	2.965	3.132	9.450	0.0991	0.1318	0.3630	146.9	573.4	208.1
54,000	390.0	216.7	2.826	2.986	9.916	0.0944	0.1256	0.3544	140.0	573.4	203.2
55,000	390.0	216.7	2.693	2.845	10.40	0.0900	0.1197	0.3460	133.4	573.4	198.4
56,000	390.0	216.7	2.567	2.712	10.92	0.0858	0.1141	0.3378	127.2	573.4	193.7
57,000	390.0	216.7	2.446	2.585	11.45	0.0818	0.1087	0.3298	121.2	573.4	189.1
58,000	390.0	216.7	2.331	2.463	12.02	0.0779	0.1036	0.3219	115.5	573.4	184.6
59,000	390.0	216.7	2.222	2.348	12.61	0.0743	0.0988	0.3143	111.0	573.4	180.2
60,000	390.0	216.7	2.118	2.238	13.23	0.0708	0.0941	0.3068	104.9	573.4	175.9
61,000	390.0	216.7	2.018	2.132	13.88	0.0675	0.0897	0.2995	99.98	573.4	171.7
62,000	390.0	216.7	1.924	2.032	14.56	0.0643	0.0855	0.2924	95.30	573.4	167.7
63,000	390.0	216.7	1.833	1.937	15.28	0.0613	0.0815	0.2855	90.84	573.4	163.7
64,000	390.0	216.7	1.747	1.846	16.04	0.0584	0.0777	0.2787	86.61	573.4	159.8
65,000	390.0	216.7	1.665	1.760	16.82	0.0557	0.0740	0.2721	82.48	573.4	156.0

Aerodynamics, continued

Airspeed Relationships

IAS —indicated airspeed (read from cockpit instrumentation, includes cockpit-instrument error correction)

CAS—calibrated airspeed (indicated airspeed corrected for airspeed-instrumentation position error)

EAS—equivalent airspeed (calibrated airspeed corrected for compressibility effects)

TAS —true airspeed (equivalent airspeed corrected for change in atmospheric density)

$$TAS = \frac{EAS}{\sqrt{\sigma}}$$

Mach number:

$$M = \frac{V_a}{a} = V_a \big/ \sqrt{\gamma g R T}$$

where

V_a = true airspeed
a = sonic velocity
γ = specific heat ratio
g = gravitational constant
R = gas constant
T = ambient temperature

Change in velocity with change in air density at constant horsepower:

$$V_2 = V_1 \sqrt[3]{\frac{\rho_1}{\rho_2}} \quad \text{(approximate)}$$

Change in velocity with change in power at constant air density:

$$V_2 = V_1 \sqrt[3]{\frac{hp_2}{hp_1}} \quad \text{(approximate)}$$

The following are equivalent at 15,000 ft, 30°C day:

M = 0.428
TAS = 290 kn
CAS = 215 kn
EAS = 213 kn

Aerodynamics, continued

Airspeed Conversion Charts

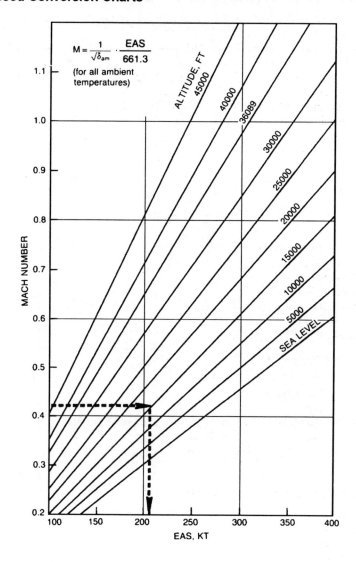

$$M = \frac{1}{\sqrt{\delta_{am}}} \cdot \frac{EAS}{661.3}$$

(for all ambient temperatures)

Aerodynamics, continued

Airspeed Conversion Charts, continued

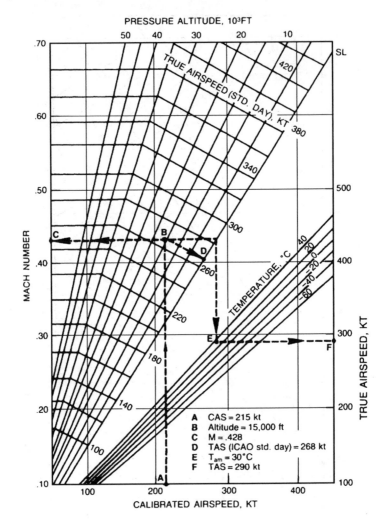

A   CAS = 215 kt
B   Altitude = 15,000 ft
C   M = .428
D   TAS (ICAO std. day) = 268 kt
E T_{am} = 30°C
F   TAS = 290 kt

Aerodynamics, continued

Airspeed Conversion Charts, continued

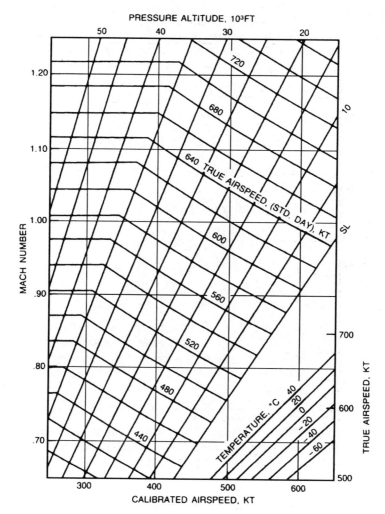

Aerodynamics, continued

Minimum Drag

Subsonic

The basic minimum drag of an aerodynamic vehicle consists of not only friction drag but also drag due to the pressure forces acting on the vehicle.

The following equations present a method of predicting minimum drag:

$$C_{D_{min}} = \frac{\text{Minimum drag}}{qS}$$

$$= \frac{\sum \left(C_{f_{comp}} \times A_{w_{comp}}\right)}{S} + C_{D_{camber}} + C_{D_{base}} + C_{D_{misc}}$$

where the component skin-friction coefficients are evaluated according to the following equations.

$C_{f_{wing}} \qquad = C_{f_{FP}}[1 + L(t/c) + 100(t/c)^4]R_{LS}$

$C_{f_{fuselage}} \qquad = C_{f_{FP}}[1 + 1.3/\overline{FR}^{1.5} + 44/\overline{FR}^3]R_{fus}$

$C_{f_{nacelle}} \qquad = C_{f_{FP}} \times Q[1 + 0.35/(\overline{FR})]$

$C_{f_{canopy}} \qquad = C_{f_{FP}}[1 + 1.3/\overline{FR}^{1.5} + 44/\overline{FR}^3]$

$C_{f_{horiz\ \&\ vert\ tails\ (one\ piece)}} = C_{f_{FP}}[1 + L(t/c) + 100(t/c)^4]R_{LS}$

$C_{f_{horiz\ \&\ vert\ tails\ (hinged)}} = (1.1)C_{f_{FP}}[1 + L(t/c) + 100(t/c)^4]R_{LS}$

$C_{f_{ext.\ store}} \qquad = C_{f_{FP}} \times Q[1 + 1.3/\overline{FR}^{1.5} + 44/\overline{FR}^3]$

$C_{D_{camber}} \qquad = 0.7(\Delta C_L^2)(S_{EXP}/S)$ (do not use for conical camber)

$C_{D_{base}} \qquad$ = base drag: a good estimate can be obtained by using a base pressure coefficient of $C_p = -0.1$. (More detailed discussion of base drag may be found in Hoerner's *Fluid Dynamic Drag*.)

$A_{w_{comp}} \qquad$ = component wetted area

$C_{D_{min}} \qquad$ = minimum drag coefficient

$C_{f_{FP}} \qquad$ = White–Christoph's flat-plate turbulent-skin-friction coefficient based on Mach number and Reynolds number (in which characteristic length of lifting surface equals exposed m.a.c.)

$c \qquad$ = lifting surface exposed streamwise m.a.c.

$\overline{FR} \qquad$ = fineness ratio

$\qquad$ = length/diameter (for closed bodies of circular cross section)

$\qquad$ = length/$\sqrt{(\text{width})(\text{height})}$ (for closed bodies of irregular cross section and for nacelles)

$\qquad = \text{length}/a\sqrt{1 + \frac{1}{2}\left(1 - \frac{a}{b}\right)\left(\frac{a}{b}\right)^2}$ (for closed bodies of elliptic cross section, where a = minor axis and b = major axis)

Aerodynamics, continued

L = thickness location parameter
= 1.2 for $(t/c)_{max}$ located at $x \geq 0.3c$
= 2.0 for $(t/c)_{max}$ located at $x < 0.3c$

$\Lambda_{max_{t/c}}$ = sweep of lifting-surface ridgeline

q = dynamic pressure

Q = interference factor
= 1.0 for nacelles and external stores mounted out of the local velocity field of the wing
= 1.25 for external stores mounted symmetrically on the wing tip
= 1.3 for nacelles and external stores if mounted in moderate proximity of the wing
= 1.5 for nacelles and external stores mounted flush to the wing (The same variation of the interference factor applies in the case of a nacelle or external store strut-mounted to or flush-mounted on the fuselage.)

R_{LS} = lifting surface factor (see Lifting Surface Correction graph)

R_{fus} = fuselage correction factor (see Fuselage Corrections graph), $R_{e_{fus}}$ based on length

S = wing gross area

S_{EXP} = exposed wing area

t = maximum thickness of section at exposed streamwise m.a.c.

Example

Calculation of uncambered-wing drag coefficient for subsonic case.

Reference wing area, $S = 1000 \text{ ft}^2$
Wetted area, $A_w = 1620 \text{ ft}^2$
Velocity, $V = 556 \text{ ft/s} = 0.54 \text{ M}$
Altitude, $H = 22{,}000 \text{ ft}$
Mean chord, $\bar{c} = 12.28 \text{ ft}$
Sweep at maximum thickness $t/c = 11\%$ at 35% chord = 24 deg
Reynolds number $= V\bar{c}/\nu = (556)(12.28)/2.7721 \times 10^{-4} = 24.63 \times 10^6$
Thickness location parameter, $L = 1.2$
Lifting surface factor, $R_{LS} = 1.13$ (see Lifting Surface Correction graph)
Basic skin-friction coefficient, $C_{f_{FP}} = 0.00255$
Wing skin-friction factor, $C_f = 0.0033$
Wing minimum-drag coefficient, $C_d = 0.00535$

Aerodynamics, continued

Subsonic-Component Correction Factors

Lifting Surface Correction

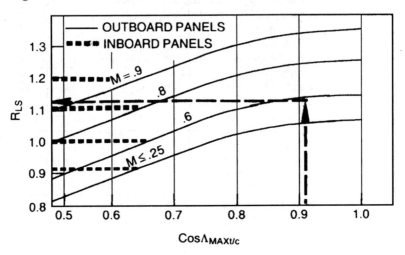

CosΛ$_{MAXt/c}$

Fuselage Corrections

Apply ratio A_{wet}/S_{ref} value for the fuselage plus attached items (to respective sets of curves, dashed or solid).

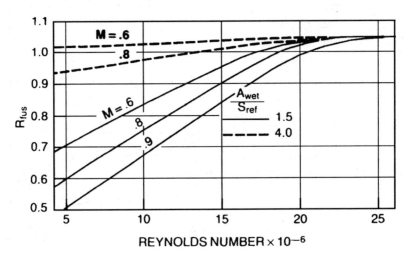

REYNOLDS NUMBER × 10^{-6}

Aerodynamics, continued

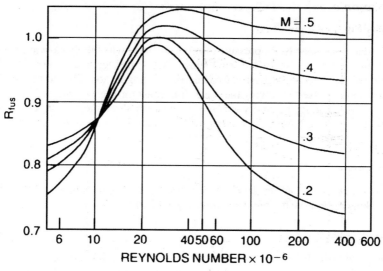

For Mach $\leq$ 0.5

Turbulent Skin-Friction Coefficient

Insulated flat plate (White–Christoph)

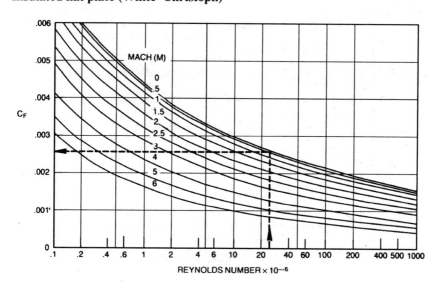

Aerodynamics, continued

Supersonic

Wave drag

At supersonic speeds, the pressure drag is associated with the shock-wave pattern about the vehicle and is called wave drag. "Area-rule" techniques are generally employed for determining the wave drag of a configuration. Due to the lengthy calculations involved in solving the wave-drag equation, digital computers are used exclusively; accuracy primarily depends on the methods employed for geometrical manipulation. The NASA Harris area-rule program is used extensively in wave-drag calculations.

Skin friction

Supersonic skin-friction drag is calculated for each component utilizing flat-plate skin-friction coefficients (see Insulated Flat Plate graph).

Wave drag—wing

As previously discussed, the calculation of wave drag for most configurations requires the resources of a digital computer. As a first approximation for wave drag of an uncambered, untwisted, conventional trapezoidal wing with sharp-nosed airfoil section, the following can be used.

$$C_{D_{\text{wave wing}}} = \frac{K_{LE}(t/c)^2}{\beta} \quad \text{(for } \beta \tan \epsilon \geq 1)$$

$$C_{D_{\text{wave wing}}} = K_{LE} \tan \epsilon (t/c)^2 \quad \text{(for } \beta \tan \epsilon < 1)$$

$$\text{(for } \epsilon = 90 \deg - \Lambda_{LE})$$

where the shape factor K_{LE} is shown in the following table.

Configuration	K_{LE}
Single wedge	1
Symmetrical double wedge	4
Double wedge with maximum thickness at x/c	$\dfrac{(c/x)}{(1 - x/c)}$
Biconvex section	5.3
Streamline foil with $x/c = 50\%$	5.5
Round-nose foil with $x/c = 30\%$	6.0
Slender elliptical airfoil section	6.5

Aerodynamics, continued

Wave drag—body

A first approximation for a body can be obtained from the preliminary wave-drag evaluation graph here for $M = 1.2$ using the expression:

$$C_{D_{\text{wave body}}} = \frac{A_{\text{MAX}}}{S_{\text{REF}}} \frac{K_{\text{BODY}}}{\overline{\text{FR}}^2}$$

Example

Calculate the wing wave-drag coefficient for the following conditions.

Mach number $= 1.2$
Airfoil $= 6\%$ thick symmetrical double wedge ($K_{\text{LE}} = 4.0$)
$\beta = 0.663$
$\epsilon = 90$ deg
$\beta \tan \epsilon = \infty$

$$C_{D_{\text{wave wing}}} = 0.0217$$

Example

Calculate the body wave-drag coefficient for the following conditions.

Mach number $= 1.2$
Fuselage $= 3$ ft diameter, 30 ft length ($\overline{\text{FR}} = 10$)
$dA/dx = 20$ ft
Reference wing area $= 67$ ft^2
$K_{\text{BODY}} = 18$ (see Preliminary Wave-Drag Evaluation graph)

$$C_{D_{\text{wave body}}} = 0.0190$$

Preliminary Wave-Drag Evaluation

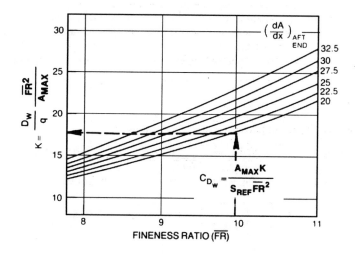

Aerodynamics, continued

Induced Drag

Subsonic

The drag due to lift, or induced drag, reflects lift-producing circulation. Potential flow theory shows that the relationship to drag is a function of lift squared. Hence, the basic polar is parabolic. The parabolic polar is shifted from the origin by camber, wing incidence, minimum drag, etc. and deviates from its parabolic shape at higher lifts when flow separation exists.

$$C_{D_i} = K C_L^2$$

where C_L is total lift, including camber effects, and K = the parabolic drag constant. For plain wings below the parabolic polar break lift coefficient,

$$K = \frac{1}{\pi(ARe)}$$

The value of e, the wing efficiency factor, accounts for the non-ellipticity of the lift distribution; typical values of e for high-subsonic jets are 0.75–0.85. The higher the wing sweep angle, the lower the e factor.

For wings with sharp leading edges, the drag due to lift can be approximated by $C_{D_i} = 0.95 C_L \tan \alpha$ (α = wing angle of attack).

Example

Induced drag at a lift coefficient of 0.8 for a vehicle with an effective Oswald efficiency factor e of 0.80 and an aspect ratio of 8.5 will be

$$C_{D_i} = \frac{C_L^2}{\pi ARe} = \frac{(0.8)^2}{\pi(8.5)(0.80)} = 0.030$$

Supersonic

At supersonic speeds, the wave drag due to lift increases as $\sqrt{M^2 - 1}$ and is a function of planform geometry. For preliminary design evaluation, the following graphs provide sufficient accuracy. At supersonic speeds, the polars generally show no tendency to depart from a parabolic shape. Thus, there is no corresponding polar break as at subsonic speeds.

Example

For straight tapered planform with sharp leading edges,

Mach number = 1.2	Wing area, $S = 67$ ft^2	$P = S/b\ell = 1.0$
Aspect ratio = 1.5	Length = 6.69 ft	$\beta = 0.663$
Span, $b = 10.02$	Taper ratio = 1.0	$\beta(b/2\ell) = 0.497$

$$\pi AR \frac{C_{D_i}}{C_L^2} \left(\frac{P}{1+P}\right) = 0.6 \qquad C_{D_i} = 0.255 C_L^2$$

Aerodynamics, continued

Supersonic Drag Due to Lift (Empirical)

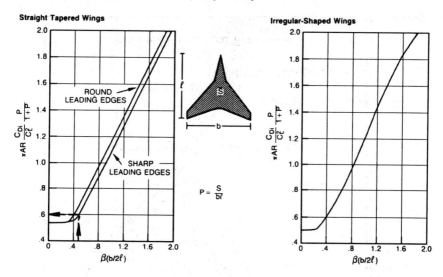

Critical Mach Number

Subsonic drag evaluation terminates at a Mach number where the onset of shock formations on a configuration causes a sudden increase in the drag level—the so-called "critical Mach number." The following graphs show simple working curves for it.

Example

Find critical Mach number for $C_L = 0.4$.

$t/c = 0.068$
$C_{L_{DESIGN}} = 0.2$
Aspect ratio $= 3$
$\Lambda_{c/4} = 45°$
$M_{CR_{CL=0}} = 0.895$ (see Critical Mach Number graph)
$M_{CR_{CL=0.4}}/M_{CR_{CL=0}} = 0.97$ (see Lift Effect on Critical Mach Number graph)

$$M_{CR_{CL=0.4}} = (0.97)(0.895) = 0.868$$

Aerodynamics, continued

Critical Mach Number

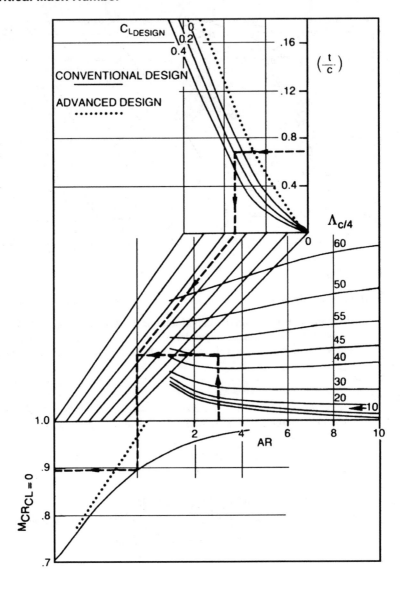

Aerodynamics, continued

Lift Effect on Critical Mach Number

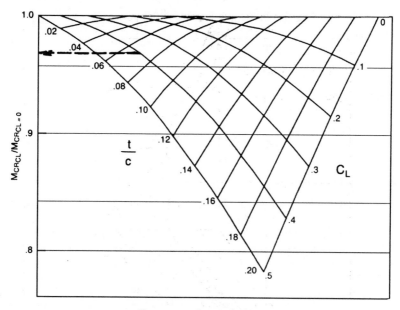

For conventional airfoils

Drag Rise

Following the critical Mach number, the drag level increases abruptly. This phenomenon is associated with strong shocks occurring on the wing or body, causing flow separation. To estimate the drag rise increment at these conditions, Hoerner, in *Fluid Dynamic Drag*, gives the following empirical relation.

$$\Delta C_{D_{\text{Rise}}} = (K/10^3)\left[10\Delta M \bigg/ \left(\frac{1}{\cos \Lambda_{\text{LE}}} - M_{\text{CR}}\right)\right]^n$$

where

K $\quad= 0.35$ for 6-series airfoils in open tunnels
$\quad\quad= 0.40$ for airfoil sections with $t/c \approx 6\%$
$\quad\quad= 0.50$ for thicker airfoils and for 6-series airfoils
$\Delta M = M - M_{\text{CR}}$
$n \quad = 3/(1 + \frac{1}{\text{AR}})$

Aerodynamics, continued

Example

Determine the drag rise increment for the following.

$\Delta M = 0.05$
Aspect ratio $= 3.0$
$t/c = 0.068 \ (K = 0.40)$
$\Lambda_{LE} = 50°$
$M_{CR} = 0.895$

$$\Delta C_{D_{Rise}} = \frac{0.4}{1000} \left[\frac{10(0.05)}{\cos(50°)^{-1} - 0.895} \right]^{\frac{3}{1+1/3}} = 0.0002136$$

Aerodynamic Center

The prediction of wing-alone aerodynamic center (a.c.) may be made from curves presented in the following graphs which show a.c. location as a fraction of the wing root chord. These curves are based on planform characteristics only and are most applicable to low-aspect-ratio wings. The characteristics of high-aspect-ratio wings are primarily determined by two-dimensional section characteristics of the wing.

The wing is the primary component determining the location of the airplane a.c., but aerodynamic effects of body, nacelles, and tail must also be considered. These effects can be taken into account by considering each component's incremental lift with its associated center of pressure and utilizing the expression,

$$\text{a.c.} = -\frac{C_{M_\alpha}}{C_{L_\alpha}}$$

Example

Determine the location of the aerodynamic center of a wing under the following conditions.

Mach number $= 1.2 \quad (\beta = 0.663)$
$\Lambda_{LE} = 45 \deg$
Aspect ratio $= 2.0$
Taper ratio $(\lambda) = 0.2$

$$\frac{X_{ac}}{C_r} = 0.51 \quad \text{(see Location of Wing Aerodynamic Center graph)}$$

Aerodynamics, continued

Location of Wing Aerodynamic Center

$\lambda = 0$

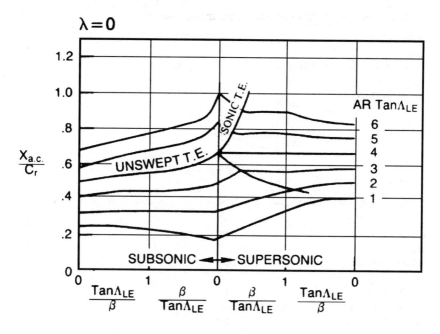

$\lambda = 0.2$

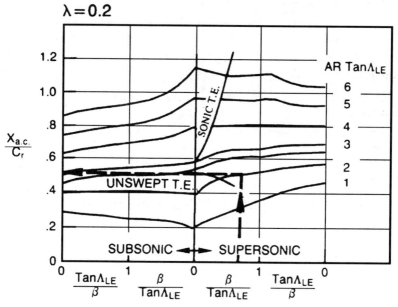

Aerodynamics, continued

Location of Wing Aerodynamic Center, continued

λ=0.25

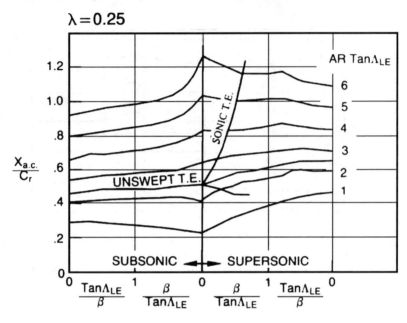

λ=0.33

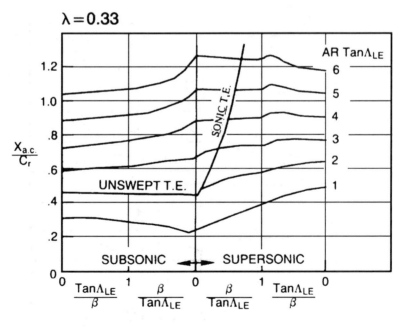

Aerodynamics, continued

Location of Wing Aerodynamic Center, continued

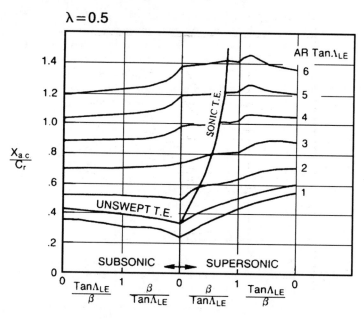

$\lambda = 0.5$

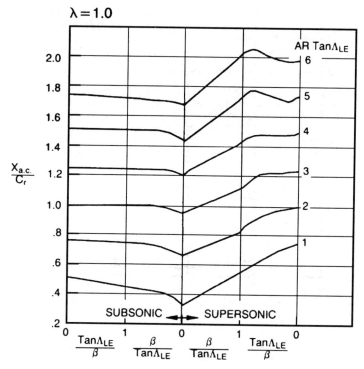

$\lambda = 1.0$

Aerodynamics, continued

Maximum Lift Coefficient (C_{Lmax})

Determination of maximum lift depends exclusively on the viscous phenomena controlling the flow over the wing. As wing incidence increases, flow separates from the surface, and lifting pressures cease to be generated. These phenomena depend on the wing geometry: sweep, aspect ratio, taper ratio, thickness ratio, and airfoil section. The thickness ratio has a decidedly marked effect through the influence of the leading-edge radius. Being a viscous phenomenon, a strong effect of Reynolds number is apparent. As a consequence, the precise determination of C_{Lmax} for an arbitrary wing has never been satisfactorily accomplished. The *USAF Stability and Control Handbook* (WADD-TR-60-261: Lib. 57211) does contain such a method. However, correlations indicate errors exceeding 25%. This method works with section lift data and applies corrections for finite airplane effects. Therefore, predictions must depend on existing test data and experience.

High-Lift Devices

Maximum Lift Increment Due to Flaps

High-lift devices are designed for certain specialized functions. Generally they are used to increase the wing camber or in some other way to control the flow over the wing, for example, to prevent flow separation. Wing flaps increase the camber at the wing trailing edge, thus inducing a higher lift due to increased circulation at the same angle of attack as the plain wing. Plain-flap effectiveness can be determined very accurately. Other devices customarily employed are slats, slots, and special leading-edge modification. Evaluation of these devices depends on the application of NASA results. British report Aero 2185 and NASA Technical Note 3911 contain prediction curves and techniques for these devices.

At supersonic speeds, high-lift devices are generally not required for flow stabilization. However, flap-type controls may be used to trim the airplane pitching moments.

The determination of maximum lift increment due to trailing-edge flap deflection uses the method of NASA TN 3911. The maximum lift increment is given by

$$\Delta C_{Lmax} = \Delta C_{\ell\,max} \frac{C_{L_a}}{C_{\ell_a}} K_c K_b$$

where $\Delta C_{\ell\,max}$ = increment of section lift coefficient due to flap deflection (see Princeton Report No. 349).

Aerodynamics, continued

$$C_{L_\alpha}/C_{\ell_\alpha} = \frac{AR}{\frac{a_o}{\pi} + \left[\left(\frac{a_o}{\pi}\right)^2 + \left(\frac{AR}{\cos \Lambda_{C/2}}\right)^2\right]^{\frac{1}{2}}}$$

where

a_o = section lift-curve slope, per radian
K_b = flap span factor (see Flap Span Factor graph)
K_c = flap chord factor (see Flap Chord Factor graph)

Care should be exercised in use of all prediction techniques for ΔC_{Lmax} due to flap deflection, because flap effectiveness is modified to a large extent by the ability of the wing-leading-edge devices to maintain attached flow.

Example

Determine ΔC_{Lmax} due to flap deflection, with the following wing characteristics.

Aspect ratio = 4.0
Section lift-curve slope = 5.73 per radian
Semichord sweep = 20 deg
Taper ratio = 0.2

$$C_{L_\alpha}/C_{\ell_\alpha} = \frac{4.0}{\frac{5.73}{\pi} + \left[\left(\frac{5.73}{\pi}\right)^2 + \left(\frac{4.0}{\cos(20)}\right)^2\right]^{\frac{1}{2}}} = 0.6197$$

For plain flaps with 50 deg deflection, 16% chord,

$\Delta C_{\ell max} = 0.80$ for a typical plain flap
Inboard span, $\eta_I = 0.15$ $K_{bi} = 0.22$ (see Flap Span Factor graph)
Outboard span, $\eta_O = 0.65$ $K_{bo} = 0.80$ (see Flap Span Factor graph)
$K_b = 0.8 - 0.22 = 0.58$
$(\alpha_\delta)_{c_\ell} = 0.5$ (see Flap Chord Factor graph)
$K_c = 1.1$ (see Flap Chord Factor graph)

$$C_{Lmax} = (0.8)(0.6197)(1.1)(0.58) = 0.3163$$

Aerodynamics, continued

Flap Chord Factor (K_c)

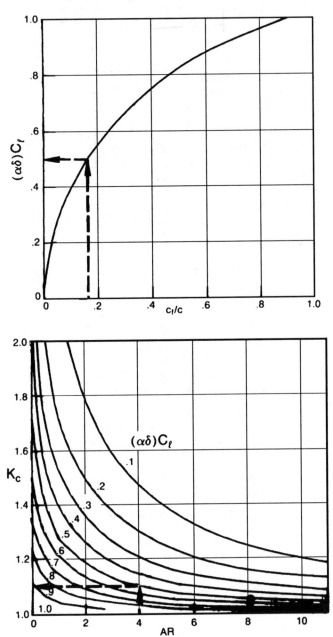

Aerodynamics, continued

Flap Span Factor (K_b)

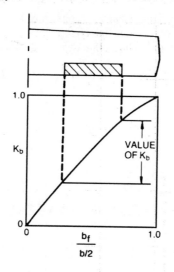

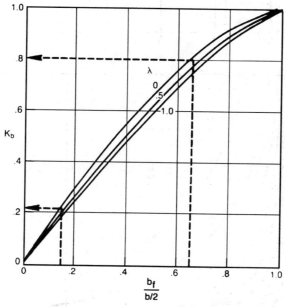

Aerodynamics, continued

Effects of Flap Deflection on Zero-Lift Angle of Attack

The set of Plain-Flap Effectiveness graphs may be used with the following expression to obtain subsonic, plain-flap effectiveness. Correction factors for body effect, partial span, and flap-leading-edge gap are shown.

$$\Delta\alpha_{Lo} = (\partial\alpha/\partial\delta)\cos\Lambda_{HL}\delta[1 - (a + b\beta)|\delta|]$$

Example

Determine $\Delta\alpha_{Lo}$ due to plain-flap deflection, with the following wing characteristics.

Aspect ratio $= 4.0$
Sweep at flap hinge line $= 10$ deg
Sweep at wing leading edge $= 32$ deg

For inboard plain flaps with 25 deg deflection, 15% flap chord,

Flap gap ratio, $GAP/\bar{c}_f = 0.002$
Inboard span, $\eta_I = 0.15$
Outboard span, $\eta_O = 0.65$
Ratio of body diameter to wing span, $2R_O/b = 0.15$
Approach Mach number $= 0.1$
$\beta = \sqrt{1 - M^2} = 0.995$
Flap effectiveness $= -\partial\alpha/\partial\delta = 0.49$ (see Plain-Flap Effectiveness graph)
   Sweep factor:

$$-a = 0.039$$
$$\qquad\qquad\text{(see Sweep factor graph)}$$
$$b = 0.034$$

Fuselage factor:

$$\frac{(\Delta\alpha_{Lo})_{WB}}{(\Delta\alpha_{Lo})_W} = 0.93 \quad\text{(see Wing–body graph)}$$

Flap span factor:

$$S_{F(PS)}/S_{F(FS)} = \eta_0 - \eta_1 = 0.65 - 0.15 = 0.50$$

$$\frac{(\Delta\alpha_{Lo})_{PS}}{(\Delta\alpha_{Lo})_{FS}} = 0.55 \quad\text{(see Part-span graph)}$$

Flap gap factor:

$$\frac{(\Delta\alpha_{Lo})_{gap}}{(\Delta\alpha_{Lo})_{sealed}} = 0.96 \quad\text{(see Gap graph)}$$

$$\Delta\alpha_{Lo} = (-0.49)\cos(10\deg)(25)[1 - (-0.039 + 0.034(0.995))(25)]$$
$$\times (0.93)(0.55)(0.96) = -6.69\deg$$

Aerodynamics, continued

Plain-Flap Effectiveness

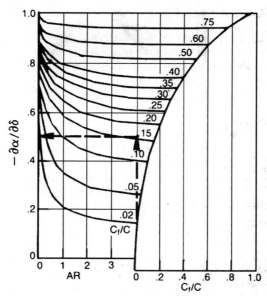

Full-span effectiveness

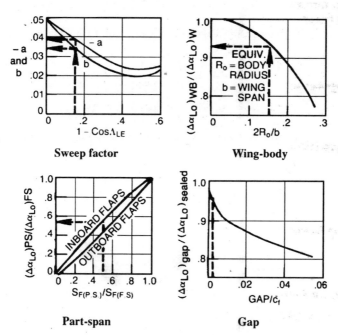

Sweep factor   **Wing-body**

Part-span   **Gap**

Aerodynamics, continued

Flaps Deployed

Induced drag due to flaps deployed

Subsonic. Trailing-edge flaps on a wing give it a camber for improving lift. Elevons in a similar manner camber the wing, although they are utilized for control at cruise lift values. Drag due to lift ratio can be expressed as follows.

$$\frac{C_{Di\delta}}{C_{Di\delta} = 0} = (1 - F)\left(\frac{\alpha - \alpha_{Lo_\delta}}{\alpha - \alpha_{Lo}}\right)^2 + F\left(\frac{\alpha - \alpha_{Lo_\delta}}{\alpha - \alpha_{Lo}}\right)$$

where $F = \sin \Lambda_{c/4}(0.295 + 0.066\alpha - 0.00165\alpha^2)$.

The following graph shows working plots of these relationships. Zero-lift angle of attack with flaps deflected is obtained from the preceding section.

Supersonic. At the supersonic speeds, flaps as such are not likely to be used. However, elevons still are required on tailless configurations for pitch control. The subsonic data listed above may be used.

Example

Find induced-drag ratio for flaps deployed for the following wing characteristics: angle of attack at zero lift α_{Lo} of -6 deg and quarter-chord sweep $\Lambda_{c/4} = 26$ deg. For a plain flap with 25 deg deflection, and with angle of attack at zero-lift with flaps deflected α_{Lo_δ} of -10 deg, the induced-drag ratio at 10 deg angle of attack will be as follows.

$$\frac{F}{\sin \Lambda_{c/4}} = 0.79 \text{ (see Drag-Due-to-Lift Ratio with Flap Deflection graph)}$$

$$\frac{\alpha - \alpha_{Lo_\delta}}{\alpha - \alpha_{Lo}} = \frac{10 - (-10)}{10 - (-6)} = 1.25$$

$$F = (0.79)\sin(26 \text{ deg}) = 0.3463$$

$$\frac{C_{Di\delta}}{C_{Di\delta} = 0} = 1.454 \text{ (see Drag-Due-to-Lift Ratio with Flap Deflection graph)}$$

Aerodynamics, continued

Drag-Due-to-Lift Ratio with Flap Deflection

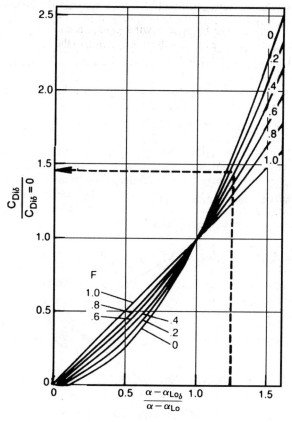

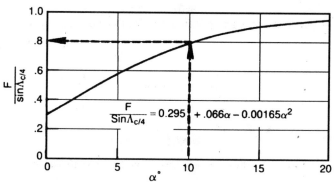

$$\frac{F}{Sin\Lambda_{c/4}} = 0.295 + .066\alpha - 0.00165\alpha^2$$

Radar Cross Section

Radar Equation

The term stealth pertaining to aircraft has been associated with invisible to radar and low radar cross section (RCS). In fact, radar is only one of several sensors that is considered in the design of a low-observable aircraft platform. Others include infrared (IR), optical (visible), and acoustic (sound) sensors. It is also important that a low-observable target have low emissions such as low radar cross section. Stealthy targets are not completely invisible to radar. To be undetectable, it is only necessary that a target's RCS be low enough for its echo return to be below the detection threshold of the radar. Radar cross section reduction has evolved as a countermeasure against radars and, conversely, more sensitive radars have evolved to detect lower RCS targets.

The radar equation describes the performance of a radar for a given set of operational, environmental, and target parameters. In the most general case, the radar transmitter and receiver can be at different locations when viewed from the target, as shown in Fig. 7.1. This is referred to as bistatic radar. In most applications, the transmitter and receiver are located on the same platform and frequently share the same antenna. In this case the radar is monostatic.

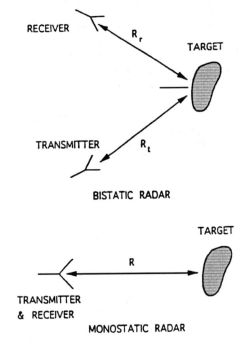

Fig. 7.1   Illustration of bistatic and monostatic radar.

Radar Cross Section, continued

The classical form of the radar equation is

$$P_r = \frac{P_t D_t D_r \sigma \lambda^2}{(4\pi)^3 R^4}$$

based on the following radar and target geometry,

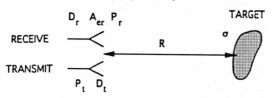

where

R = distance between monostatic radar and target
P_t = radar transmitting power
P_r = radar receive power
D_t = the antenna directivity
D_r = the receive directivity
σ = length squared
λ = the wavelength
A_{er} = the effective operative of the receive antenna

The definition of RCS can be stated as:

$$\frac{\text{Power reflected to receiver per unit solid angle}}{\text{Incident power density}/4\pi}$$

The unit of RCS most commonly used is decibels relative to a square meter (dBsm):

$$\sigma, \text{ dBsm} = 10 \log(\sigma, \text{m}^2)$$

Radar Cross Section, continued

As indicated in Fig. 7.2, typical values of RCS range from 40 dBsm (10,000 m^2) for ships and large bombers to −30 dBsm (0.01 m^2) for insects. Modern radars are capable of detecting flocks of birds and even swarms of insects. The radar's computer will examine all detections and can discard these targets on the basis of their velocity and trajectory. However, the fact that such low RCS targets are detectable implies that the RCS design engineers have a difficult job.

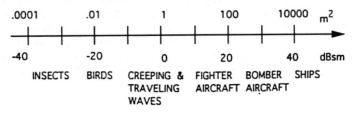

Fig. 7.2　Typical values of RCS.

A few long-range ballistic missile defense radars operate in the 300-MHz region (UHF), but most others use frequencies greater than 1 GHz (L band and above) as shown in Table 7.1.

Table 7.1　Radar frequency bands

Band designation	Frequency range
HF	3–30 MHz
VHF	30–300 MHz
UHF	300–1000 MHz
L	1–2 GHz
S	2–4 GHz
C	4–8 GHz
X	8–12 GHz
Ku	12–18 GHz
K	18–27 GHz
Ka	27–40 GHz
MM	40–300 GHz

Low frequencies are capable of handling more power because the applied voltages can be higher without causing breakdown. Finally, ambient noise is lowest in the 1–10-GHz range, and low-altitude atmospheric attenuation favors frequencies below 18 GHz.

Radar Cross Section, continued

Scattering Mechanisms

The RCS of targets encountered by most radars are more complicated than spheres or plates. There are a few exceptions, such as weather balloons and buoys. Simple shapes, however, such as plates, spheres, cylinders, and wires, are useful in studying the phenomenology of RCS. Furthermore, complex targets can be decomposed into primitives (basic geometrical shapes that can be assembled to form a more complex shape). As shown in Fig. 7.3, an aircraft can be decomposed into cylinders, plates, cones, and hemispheres. A collection of basic shapes will give an acceptable RCS estimate that can be used during the initial design stages of a platform. The locations and levels of the largest RCS lobes are of most concern at this stage of the design process. The accuracy of the RCS calculation at other angles will depend on how the interactions between the various shapes are handled.

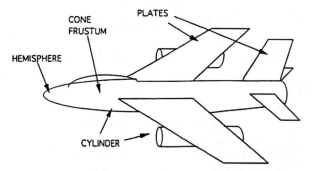

Fig. 7.3 Aircraft represented by geometrical components.

Several scattering mechanisms are commonly encountered. Their level relative to the peak RCS may be quite small, but, away from the peaks, scattering due to these mechanisms can dominate. They are depicted in Fig. 7.4 and are briefly described as follows.

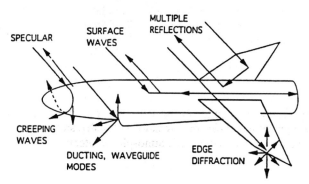

Fig. 7.4 Illustration of scattering mechanisms.

Radar Cross Section, continued

Reflections. This mechanism yields the highest RCS peaks, but these peaks are limited in number because Snell's law must be satisfied. When multiple surfaces are present, multiple reflections can occur. For instance, the incident plan wave could possibly reflect off the fuselage, hit a fin, and then return to the radar.

Diffractions. Diffracted fields are those scattered from discontinuities such as edges and tips. The waves diffracted from these shapes are less intense than reflected waves, but they can emerge over a wide range of angles. In regions of low RCS, diffracted waves can be significant.

Surface Waves. The term surface wave refers to the current traveling along a body and includes several types of waves. In general, the target acts as a transmission line, guiding the wave along its surface. If the surface is a smooth closed shape such as a sphere, the wave will circulate around the body many times. On curved bodies, the surface wave will continuously radiate. These are called creeping waves because they appear to creep around the back of a curved body. Radiating surface waves on flat bodies are usually called leaky waves. Traveling waves appear on slender bodies and along edges and suffer little attenuation as they propagate. If the surface is terminated with a discontinuity such as an edge, the traveling wave will be reflected back toward its origin. Traveling wave RCS lobes can achieve surprisingly large levels.

Ducting. (Also called waveguide modes.) Ducting occurs when a wave is trapped in a partially closed structure. An example is an air inlet cavity on a jet. Once the wave enters the cavity, many bounces can occur before a ray emerges. The ray can take many paths, and, therefore, rays will emerge at most all angles. The result is a large, broad RCS lobe. An optical analogy of this is the glowing of a cat's eye when it is illuminated by a light.

Sometimes these mechanisms interact with each other. For example, a wave reflected from a flat surface can subsequently be diffracted from an edge or enter a cavity. For a complex target, the interactions are not always obvious.

Radar Cross Section Design Guidelines

For any system or platform, RCS reduction is achieved at the expense of almost every other performance measure. In the case of an aircraft, stealth has resulted in decreased aerodynamic performance and increased complexity. Complexity, of course, translates into increased cost. The trend in the design of new platforms has been to integrate all the engineering disciplines into a common process. Thermal, mechanical, aerodynamic, and RCS simulations are carried out in parallel, using a common database. Changes made to the structure or materials for the purpose of decreasing RCS are automatically recorded in the databases for all other design simulations. This approach is called concurrent engineering.

Radar Cross Section, continued

Treating the most intense sources of scattering is the easiest, and the payoff in decibels is greatest. For example, just tilting a surface a few degrees can reduce the RCS presented to a monostatic radar by 20 or 30 dB. Further reduction is much more difficult because second-order scattering mechanisms (multiple reflections, diffractions, surface waves, etc.) become important. Therefore, it can be more costly to drop the RCS the next 5 dB than it was for the first 30 dB. With that point in mind, it is evident that the guidelines for designing a low-RCS vehicle will not be as extensive as those for an ultralow-RCS vehicle. The guidelines for both vehicles have several basic points in common, however, and these are summarized as follows.

1) Design for specific threats when possible to minimize cost. Keep in mind the threat radar frequency, whether it is monostatic or bistatic, and the target aspect angles that will be presented to the radar.
2) Orient large, flat surfaces away from high priority quiet zones.
3) Use lossy materials or coatings to reduce specular/traveling wave reflections.
4) Maintain tolerances on large surfaces and materials.
5) Treat trailing edges to avoid traveling wave lobes.
6) Avoid corner reflectors (dihedrals or trihedrals).
7) Do not expose cavity inlets; use a mesh cover or locate the inlets out of view of the radar.
8) Shield high-gain antennas from out-of-band threats.
9) Avoid discontinuities in geometry and materials to minimize diffraction and traveling wave radiation.

As an example of the application of these points, consider Figs. 7.5 and 7.6. Figure 7.5 shows a sketch of a typical fighter aircraft with some scattering sources labeled. The relative strength of each scatterer is also shown. Figure 7.6 shows a more stealthy design that has incorporated many of the points just listed.

Radar Cross Section, continued

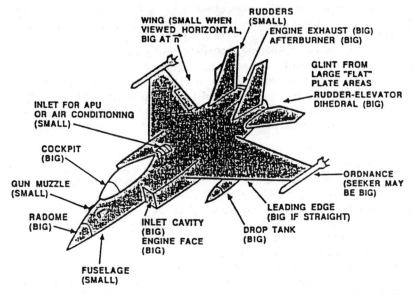

Fig. 7.5 Standard jet fighter aircraft design with scattering sources indicated. (Ref.: Fuhs, A. E., *Radar Cross Section Lectures*, AIAA, New York, 1984.)

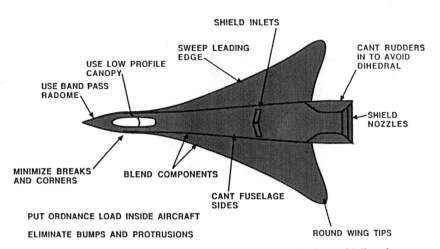

Fig. 7.6 Stealthy jet fighter aircraft design with RCS reduction guidelines incorporated. (Ref.: Fuhs, A. E., *Radar Cross Section Lectures*, AIAA, New York, 1984.)

Propulsion

Propeller Propulsion Nomenclature

$C_p = \text{PW}/\rho n^3 D^5 = \text{power coefficient}$

$C_s = \sqrt[5]{\rho V^5/\text{PW}n^2} = \text{speed power coefficient}$

$C_T = F/\rho n^2 D^4 = \text{thrust coefficient}$

$C_Q = Q/\rho n^2 D^5 = \text{torque coefficient}$

$J = V/nD = \text{advance ratio}$

$\eta = FV/\text{PW} = \text{propeller efficiency}$

$F_{\text{hp}} = FV/550 = \text{thrust horsepower}$

where

D = propeller diameter, ft

n = propeller speed, rps

PW = power, ft-lb/s

Q = torque, lb-ft

F = thrust, lb

V = aircraft speed, fps

ρ = air density at altitude, lb-s^2/ft^4

Propulsion, continued

Propeller Thrust

Propeller Thrust per Shaft Horsepower

As function of airspeed and propeller efficiency: $F/PW = (550/V)\eta$

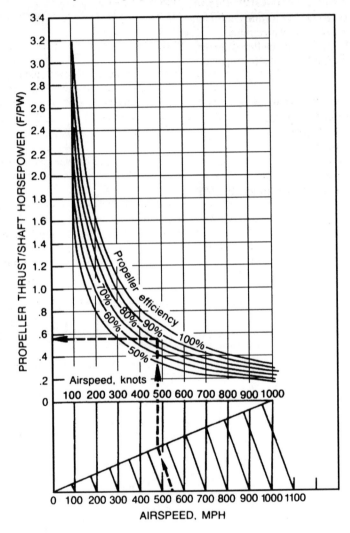

Example

$V = 550 \text{ mph}$ $\qquad$ $\eta = 80\%$ $\qquad$ $F/PW = 0.55$

Propulsion, continued

Jet Propulsion Nomenclature

Gas turbine = an engine consisting of a compressor, burner or heat exchanger, and turbine, using a gaseous fluid as the working medium, producing either shaft horsepower or jet thrust, or both

Turboprop = a type of gas turbine that converts heat energy into propeller shaft work and some jet thrust

Turbojet = a gas turbine whose entire propulsive output is delivered by the jet

Pulse-jet = a jet engine whose thrust impulses are intermittent; usually simple duct with some type of air-control valves at the front end

Ramjet (Athodyd) = a jet engine consisting of a duct utilizing the dynamic head (due to the forward motion) for compression of the air and then producing thrust by burning fuel in the duct and expanding the result through a nozzle

Rocket motor = a jet engine that carries its entire supply of fuel and oxidizer and exhausts their combustion products through a nozzle to produce thrust

Open cycle = a thermodynamic cycle in which the working fluid passes through the system only once, being thereafter discharged

Closed cycle = a thermodynamic cycle in which the working fluid recirculates through the system

Regenerator = a heat exchanger that transfers heat from the exhaust gas to the fluid after compression, before the burner, to increase cycle thermal efficiency

Reheater = a burner (or heat exchanger) that adds heat to the fluid, between turbine stages, to increase the power from a given size of machine

Intercooler = a heat exchanger that cools the fluid between stages of compression to decrease the work of compression

Afterburner = a burner that adds heat to the fluid after the turbine stages to increase the thrust from a given size of machine

Propulsion, continued

Gas Turbine and Turbojet Performance Parameters

Corrected to engine-inlet total conditions.

Quantity	Symbol	Parameter
Thrust	$F(F_g, F_n, \text{etc.})$	$\dfrac{F}{\delta}$
Fuel flow	W_f	$\dfrac{W_f}{\delta\sqrt{\theta}}$
Air flow	W_a	$\dfrac{W_a\sqrt{\theta}}{\delta}$
Fuel/air ratio	W_f/W_a	$\dfrac{W_f}{W_a\theta}$
rpm	N	$\dfrac{N}{\sqrt{\theta}}$
Specific fuel consumption	W_f/F	$\dfrac{W_f}{F\sqrt{\theta}}$
Specific air consumption	W_a/F	$\dfrac{W_a\sqrt{\theta}}{F}$
Any absolute temperature	T	$\dfrac{T}{\theta}$
Any absolute pressure	P	$\dfrac{P}{\delta}$
Any velocity	$V(\text{or } v \text{ or } u)$	$\dfrac{V}{\sqrt{\theta}}$

Propulsion, continued

Gas-Turbine Correction Parameters

Corrected net thrust:

$$F_{n_{(corr)}} = \frac{F_{n(observed)}}{\delta}$$

Corrected fuel flow:

$$W_{f_{(corr)}} = \frac{W_{f(observed)}}{\delta \sqrt{\theta}^{\,a}}$$

Corrected thrust specific fuel consumption:

$$TSFC_{(corr)} = \frac{W_{f_{(corr)}}}{F_{n_{(corr)}}} = \frac{TSFC}{\sqrt{\theta}^{\,a}}$$

Corrected air flow:

$$W_{a_{(corr)}} = \frac{W_{a(observed)} \sqrt{\theta}}{\delta}$$

Corrected exhaust gas temperature:

$$EGT_{(corr)} = \frac{EGT_{(observed)}}{\theta^a}$$

Corrected rotor speed:

$$N_{(corr)} = \frac{N_{(observed)}}{\sqrt{\theta}}$$

Corrected shaft power:

$$PW_{(corr)} = \frac{PW_{observed}}{\delta \sqrt{\theta}^{\,a}}$$

$$\theta = \frac{T}{T_o} \qquad \delta = \frac{P}{P_o}$$

[a]The exponent for θ is a function of the engine cycle as developed from theoretical and empirical data. The exponent is approximately 0.5 for correcting fuel flow and TSFC and approximately 1.0 for correcting temperature.

Propulsion, continued

Gas Turbine Station Notations

Single-Spool Turbojet/Turboshaft

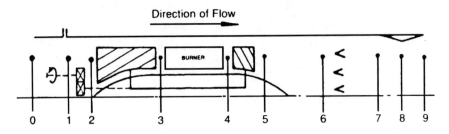

Twin-Spool Turbofan

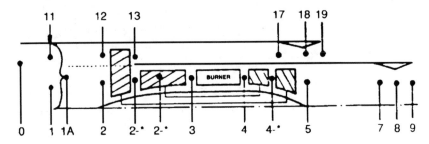

Mixed Twin-Spool Turbofan

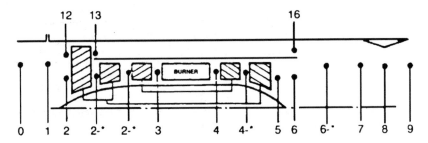

Propulsion, continued

Triple-Spool Turbofan

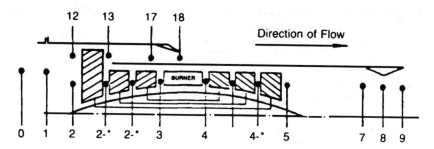

Twin-Spool Duct Heater

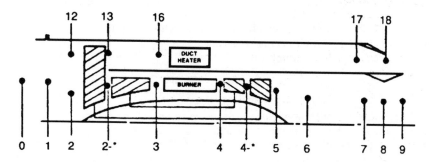

Free-Turbine Turboprop, Turboshaft

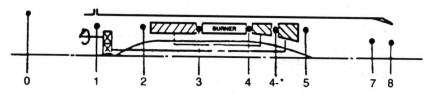

Ref.: for gas turbine station notations appearing on pages 7-50 and 7-51: SAE Aerospace Recommended Practice ARP 775A.

Performance

Nomenclature

A = margin above stall speed (typically 1.2 for takeoff and 1.15 for landing)
a = speed of sound at altitude
AR = aspect ratio
C_D = drag coefficient
C_L = lift coefficient
D = $(C_D)(q)(S)$ = aircraft drag
e = Oswald efficiency factor (wing)
F_{pl} = planform factor for takeoff calculations
F = Thrust
F/D = ratio of thrust to drag
ΔFuel = fuel increment
g = acceleration due to gravity
h = altitude
h_{obs} = height of obstacle
K = $1/(\pi)(AR)(e)$ = parabolic drag polar factor
k = $(L/D)(V/SFC)$ = range constant
L = $(C_L)(q)(S)$ = aircraft lift
L/D = ratio of lift to drag
M = Mach number
M_{DD} = drag divergence Mach number
n = load factor (lift/weight)
P = atmospheric pressure at altitude
P_s = specific excess power
q = $\frac{1}{2}\rho V^2$ = dynamic pressure
R = range
R/C = rate of climb
R/D = rate of descent
S = reference wing area
S_{air} = air run distance
S_{gnd} = ground roll distance
S_{land} = total landing distance
S_{obs} = distance to clear obstacle
SFC = W_f/F = specific fuel consumption
S_{TO} = total takeoff distance
SR = V/W_f = specific range
ΔS = distance increment
Δt = time increment
V = velocity
V_{obs} = velocity at obstacle

Performance, continued

$V_{\text{stall}} = (2W/S\rho C_{L\max})^{1/2} = \text{stall speed}$
$V_{\text{TD}} = A \times V_{\text{stall}} = \text{landing velocity}$
$\Delta V = \text{velocity increment}$
$W = \text{weight}$
$W_f = \text{fuel flow}$
$\gamma = \text{flight-path angle}$
$\mu = \text{coefficient of rolling friction}$
$\mu_{\text{BRK}} = \text{coefficient of braking friction}$
$\rho = \text{air density at an altitude}$
$\dot{\psi} = \text{rate of turn}$

Takeoff

Takeoff-distance calculations treat ground roll and the distance to clear an obstacle. Obstacle requirements differ for commercial (35 ft) and military (50 ft) aircraft.

Takeoff Ground Roll

$$S_{\text{gnd}} = -\frac{W/S}{g\rho(C_D - \mu C_L)} \ell_n \left[1 - \frac{A^2(C_D - \mu C_L)}{((F/W) - \mu)C_{L_{\max}}} \right]$$

The stall margin A typically is 1.2.

Total Takeoff Distance

$$S_{\text{TO}} = (S_{\text{gnd}})(F_{\text{pl}})$$

The factor to clear an obstacle depends greatly on available excess thrust, flight path, and pilot technique. The following typical factors characterize planforms in ability to clear a 50-ft obstacle.

Planform	F_{pl}
Straight wing	1.15
Swept wing	1.36
Delta wing	1.58

Performance, continued

Example

Takeoff distance for a straight-wing aircraft with the following characteristics:

$W = 22{,}096$ lb　　　$S_{\text{gnd}} = 1255$ ft　　　$A = 1.2$
$S = 262$ ft^2　　　　　$C_{L_{\max}} = 1.8$　　　$S_{\text{TO}} = 1444$ ft
$C_L = 0.46$　　　　　　$F = 19{,}290$　　　　$\rho = 0.0023769$ lb-s^2-ft^{-4}
$C_D = 0.3538$　　　　　$\mu = 0.025$

Takeoff Fuel Allowance

For brake release to initial climb speed,

$$\text{Fuel} = \frac{W_1 V_1}{g(F_1 - D_1)} \left(\frac{W_{f_0} \cdot W_{f_1}}{2} \right)$$

where

W_1　= weight at start of climb
V_1　= initial climb speed, ft/s
g　　= 32.174 ft/s^2
F_1　= maximum power thrust at initial climb speed
D_1　= drag at 1-g flight condition, initial climb speed
W_{f_0} = maximum power fuel flow at brake release, lb/s
W_{f_1} = maximum power fuel flow at initial climb speed, lb/s

Climb

Time, fuel, and distance to climb from one altitude (h_1) to another (h_2) can be calculated in increments and then summed. By using this technique, specific climb speed schedules—i.e., constant Mach number climb and maximum rate of climb—can be depicted.

Rate of Climb

For small angles, the rate of climb can be determined from

$$R/C = (F - D)V \Big/ W \left(1 + \frac{V}{g} \cdot \frac{dV}{dh} \right)$$

where $V/g \cdot dV/dh$ is the correction term for flight acceleration.

Performance, continued

The following table gives acceleration corrections for typical flight procedures.

Altitude, ft	Procedure	$\dfrac{V}{g} \cdot \dfrac{dV}{dh}$
36,089 or less	Constant CAS	$0.5668\ M^2$
	Constant Mach	$-0.1332\ M^2$
Over 36,089	Constant CAS	$0.7\ M^2$
	Constant Mach	Zero

For low subsonic climb speeds, the acceleration term is usually neglected.

$$R/C = (F - D)V/W$$

Flight-Path Gradient

$$\gamma = \sin^{-1}\left(\frac{F - D}{W}\right)$$

Time to Climb

$$\Delta t = \frac{2(h_2 - h_1)}{(R/C)_1 + (R/C)_2}$$

Distance to Climb

$$\Delta S = V(\Delta t)$$

Fuel to Climb

$$\Delta \text{Fuel} = W_f(\Delta t)$$

Sum increments for total.

Performance, continued

Acceleration

The time, fuel, and distance for acceleration at a constant altitude from one speed to another can be calculated in increments and then summed up. By using this technique, specific functions can be simulated (e.g., engine power spool-up).

Time-to-Accelerate Increment

$$\Delta t = \frac{\Delta V}{g\left(\frac{F-D}{W}\right)}$$

Distance-to-Accelerate Increment

$$\Delta S = V(\Delta T)$$

Fuel-to-Accelerate Increment

$$\Delta \text{Fuel} = W_f(\Delta t)$$

Sum increments to yield total time, fuel, and distance to accelerate.

Cruise

The basic cruise distance can be determined by using the Breguet range equation for jet aircraft, as follows.

Cruise Range

$$R = L/D(V/\text{SFC})\,\ell n\,(W_0/W_1)$$

where subscripts "0" and "1" stand for initial and final weight, respectively.

Cruise Fuel

$$\text{Fuel} = W_0 - W_1 = W_f\left(e^{R/k} - 1\right)$$

where k, the range constant, equals $L/D(V/\text{SFC})$.

Dash Range

For flight at constant Mach number and constant altitude, the following equation gives dash range.

$$R = \left(\frac{V}{\text{SFC}}\right)\left(\frac{1}{F}\right)(\text{Fuel})$$

Performance, continued

For large excursions of weight, speed, and altitude during cruise, it is recommended that the range calculations be divided into increments and summed up for the total.

Example

Find cruise distance for an aircraft with the following characteristics.

$W_0 = 15,800$ lb
$W_1 = 14,600$ lb
$V = 268$ kn
SFC $= 1.26$ lb/h/lb
$L/D = 9.7$

$$R = 9.7(268/1.26) \, \ell_n \, (15,800/14,600) = 163 \, \text{n mile}$$

Cruise Speeds

Cruise-speed schedules for subsonic flight can be determined by the following expressions.

Optimum Mach Number (M_{DD}), Optimum-Altitude Cruise

First calculate the pressure at altitude.

$$P = \frac{W}{0.7(M_{DD}^2)(C_{L_{DD}})S}$$

Then enter value from Cruise-Altitude Determination graph for cruise altitude.

Optimum Mach Number, Constant-Altitude Cruise

Optimum occurs at maximum $M(L/D)$.

$$M = \sqrt{\frac{W/S}{0.7P}\sqrt{\frac{3K}{C_{D_{min}}}}}$$

Constant Mach Number, Optimum-Altitude Cruise

Derive optimum altitude, as above, except M_{DD} and $C_{L_{DD}}$ are replaced with values for the specified cruise conditions.

Performance, continued

Cruise-Altitude Determination

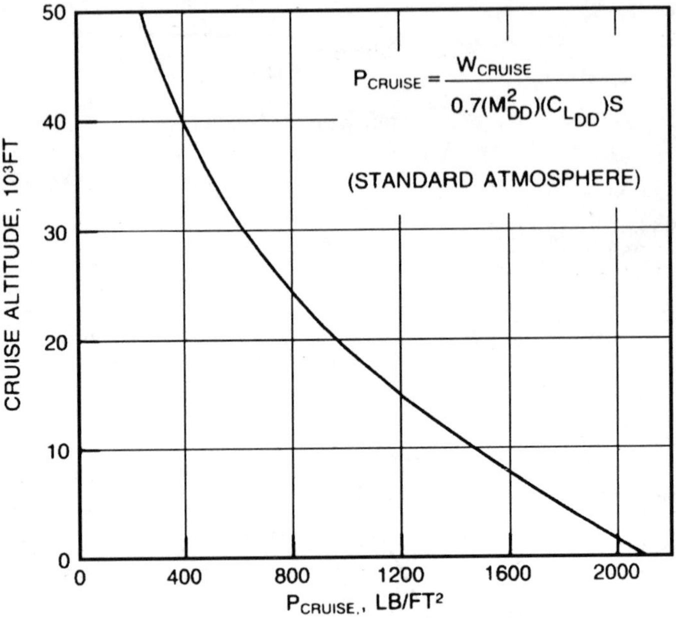

$P_{CRUISE} = \frac{W_{CRUISE}}{0.7(M^2_{DD})(C_{L_{DD}})S}$
(STANDARD ATMOSPHERE)
CRUISE ALTITUDE, 10³FT
50
40
30
20
10
0
0
400
800
1200
1600
2000
P_{CRUISE}, LB/FT²

Loiter

Loiter performance is based on conditions at $(L/D)_{\max}$, because maximum endurance is of primary concern.

Loiter Speed

$$M = \sqrt{\frac{W/S}{0.7(P)(C_L)_{(L/D)_{\max}}}}$$

where

$$C_{L(L/D)_{\max}} = \sqrt{\frac{C_{D_{\min}}}{K}}$$

Loiter Time

$$t = L/D_{\max}\left(\frac{1}{\text{SFC}}\right)\ln\left(\frac{W_0}{W_1}\right)$$

Performance, continued

Loiter Fuel

$$\text{Fuel} = W_f e^y \quad \text{where } y = \left\{ \frac{t(\text{SFC})}{L/D_{\max} - 1} \right\}$$

Example

Find time to loiter at altitude for aircraft with the following characteristics.

Initial weight, $W_0 = 11{,}074$ lb
Final weight, $W_1 = 10{,}000$ lb
$L/D_{\max} = 10.2$
SFC $= 2.175$ lb/h/lb
$t = 0.478$ h

Maneuver

The measure of maneuverability of a vehicle can be expressed in terms of sustained and instantaneous performance. For sustained performance, thrust available from the engines must equal the drag of the vehicle (i.e., specific excess power equals zero).

Specific Excess Power

$$P_S = \left(\frac{F - D}{W} \right) V$$

Turn Radius

$$\text{TR} = \frac{V^2}{g\sqrt{n^2 - 1}}$$

Turn Rate

$$\dot{\psi} = \frac{V}{\text{TR}} = \frac{g}{V}\sqrt{n^2 - 1}$$

Fuel Required for N Sustained Turns

$$\text{Fuel} = \frac{(\text{SFC})(F)(N)(\text{fac})}{\dot{\psi}}$$

where fac $= 360$ if $\dot{\psi}$ is in degrees per second; fac $= 2\pi$ if $\dot{\psi}$ is in radians per second.

Performance, continued

Example

Determine turn radius for the following vehicle.

$V = 610$ fps
$n = 1.3\ g$

$$TR = 13{,}923\ \text{ft}$$

Landing

Landing distance calculations cover distance from obstacle height to touchdown and ground roll from touchdown to a complete stop.

Approach Distance

$$S_{\text{air}} = \left(\frac{V_{\text{obs}}^2 - V_{\text{TD}}^2}{2g} + h_{\text{obs}} \right)(L/D)$$

where V_{obs} = speed at obstacle, V_{TD} = speed at touchdown, h_{obs} = obstacle height, and L/D = lift-to-drag ratio.

Landing Ground-Roll

$$S_{\text{gnd}} = \frac{(W/S)}{g\rho(C_D - \mu_{\text{BRK}}C_L)}\ \ell_n\left[1 - \frac{A^2(C_D - \mu_{\text{BRK}}C_L)}{((F/W) - \mu_{\text{BRK}})C_{L_{\max}}} \right]$$

This equation, of course, also describes takeoff roll, except being positive to account for deceleration from touchdown to zero velocity.

Normally the distance would require a two-second delay to cover the time required to achieve full braking. Commercial requirements may also dictate conservative factors be applied to the calculated distances.

Example

Find landing distance for an aircraft with the following characteristics.

W	= 17,000 lb	μ_{BRK}	= 0.6
S	= 262 ft²	A	= 1.15
C_L	= 0.46	V_{obs}	= 226 ft/s
C_D	= 0.352	V_{TD}	= 200 ft/s
$C_{L_{\max}}$	= 1.44	L/D	= 1.88
ρ	= 0.0023769 lb-s²-ft⁻⁴	h_{obs}	= 50 ft
F	= 0.0		

$$S_{\text{air}} = 418\ \text{ft} \qquad S_{\text{gnd}} = 1229\ \text{ft} \qquad S_{\text{land}} = S_{\text{air}} + S_{\text{gnd}} = 1647\ \text{ft}$$

Helicopter Design

Geometry

Positive Sign and Vector Conventions for Forces Acting on the Helicopter

Helicopter geometry is similar to that of the aircraft geometry shown on page 7-6, except for the addition of rotating blades. The following figure shows a typical side view of a helicopter with appropriate nomenclature. For a more detailed geometry breakdown, refer to the helicopter references at the end of this section.

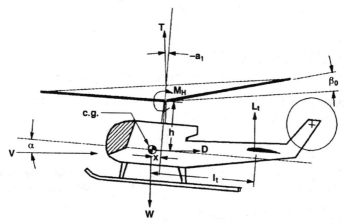

(Ref.: *Engineering Design Handbook, Helicopter Engineering, Part One: Preliminary Design*, Headquarters, U.S. Army Materiel Command, Aug. 1974.)

c.g. = center of gravity	l_t = distance, c.g. to horizontal tail
V = airflow	α = fuselage angle of incidence
x = distance, c.g. to T	$-a_1$ = mast angle
β_0 = coning angle	D = drag
M_H = hub moment	T = rotor thrust vector
L_T = lift from horizontal tail	

Preliminary Design Process

Common design requirements	Design constraints
Payload	Compliance with applicable safety standards
Range of endurance	Maximum disc loading
Critical hover or vertical climb condition	Choice of engine from list of approved engines
Maximum speed	Maximum physical size
Maximum maneuver load factor	Maximum noise level
	Minimum one-engine-out performance
	Minimum auto-rotate landing capability

Helicopter Design, continued

Outline of Typical Design Process Steps

1) Guess at the gross weight (GW in lb) and installed power ($hp_{installed}$) on the basis of existing helicopters with performance similar to that desired.

2) Estimate the fuel required (lb) using a specific fuel consumption of 0.5 for piston engines or 0.4 for turboshaft engines applied to the installed power. Estimate required mission time considering pre-takeoff warm-up, climb to cruise, cruise, descent, and operation at landing site times. Thirty minutes reserve fuel is a standard condition.

$$\text{Fuel} = \text{SFC} \times hp_{installed} \times \text{Mission time}$$

3) Calculate the useful load (UL in lb).

$$\text{UL} = \text{crew} + \text{payload} + \text{fuel (weight in lb)}$$

4) Assume a value of the ratio UL/GW based on existing helicopter and trends. Use the following figure for guidance.

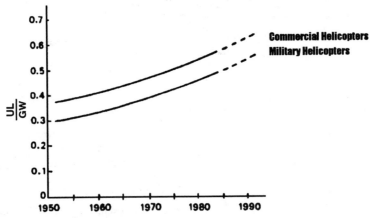

Historic trend of ratio of useful load to gross weight. (Ref.: Prouty, R. W., *Military Helicopter Design Technology*, Jane's Defence Data, UK, 1989. Reproduced with permission from Jane's Information Group.)

5) Estimate gross weight as:

$$\text{GW} = \frac{\text{UL}}{\text{UL/GW}}$$

Compare this value with the original estimate (step 1). Modify the estimate of installed power and fuel if the two calculations are significantly different.

Helicopter Design, continued

6) Assume a disc loading at the maximum allowable value, or at the highest deemed practical, and lay out the configuration based on the rotor radius corresponding to this disc loading and the estimated gross weight. The efficiency of a rotor in hover is defined as a figure of merit FM on the basis of minimum power required to hover. An FM = 0.8 represents a good state-of-the-art rotor designed primarily for hover, and the line for FM = 0.6 represents a rotor that has also been designed to give good performance at high speed resulting in a performance compromise at hover.

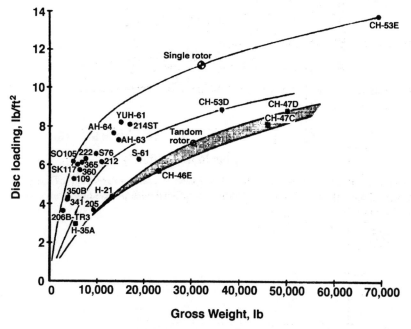

Disc loading trends. (Ref.: Rosenstein and Stanzione, "Computer-Aided Helicopter Design," AHS 37th Forum, 1981. Reproduced with permission from the American Helicopter Society.)

Helicopter Design, continued

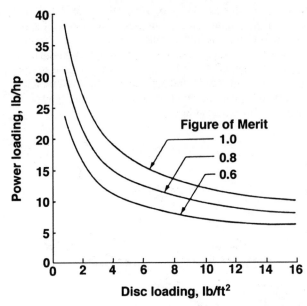

Main-rotor hover performance. (Ref.: Prouty, R. W., *Military Helicopter Design Technology*, Jane's Defence Data, UK, 1989. Reproduced with permission from Raymond W. Prouty.)

Helicopter Design, continued

7) Make first design decisions for main rotor tip speed, based on maximum speed.

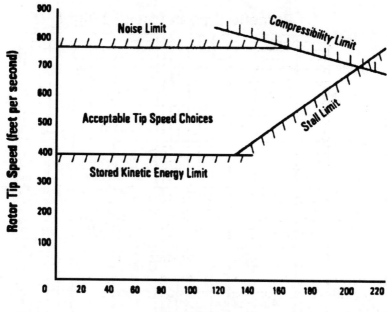

Constraints on choice of tip speeds. (Ref.: Prouty, R. W., *Military Helicopter Design Technology*, Jane's Defence Data, UK, 1989. Reproduced with permission from Raymond W. Prouty.)

Helicopter Design, continued

8) Make first estimates of drag in forward flight.

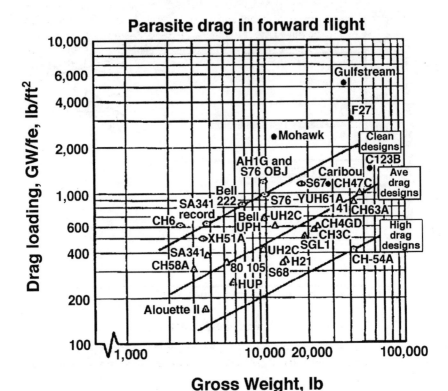

Helicopter and airplane drag state of the art. (Ref.: Rosenstein and Stanzione, "Computer-Aided Helicopter Design," AHS 37th Forum, 1981. Reproduced with permission from the American Helicopter Society.)

Helicopter Design, continued

9) Calculate installed power required to satisfy vertical rate of climb and maximum speed requirements at specified altitude and temperature. Calculating the installed power properly requires additional information such as gross weight, maximum speed, altitude, and vertical rate of climb. For a detailed discourse on installed power, refer to other sources such as *Helicopter Performance, Stability, and Control* by R. W. Prouty. Flat plate drag estimates are useful to make this calculation, and the following figure may be used.

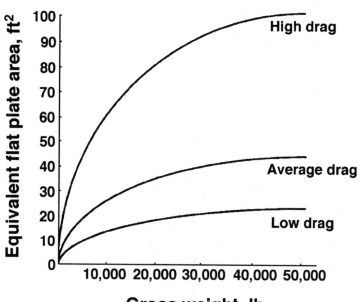

Statistical trend for equivalent flat plate area. (Ref.: Prouty, R. W., *Military Helicopter Design Technology*, Jane's Defence Data, UK, 1989. Reproduced with permission from Jane's Information Group.)

10) Select engine, or engines, that satisfy both requirements and constraints. Decrease disc loading if necessary to use approved engine if the vertical rate of climb is the critical flight condition. Use retracting landing gear and other drag reduction schemes to use approved engine if high speed is the critical flight condition.

11) Recalculate the fuel required based on the design mission and on known engine characteristics.

12) Calculate group weights based on statistical methods modified by suitable state-of-the-art assumptions. If the resulting gross weight is different from the gross weight currently being used, return to the appropriate previous step.

Helicopter Design, continued

13) Perform tradeoff studies with respect to disc loading, tip speed, solidity, twist, taper, type and number of engines, and so on to establish smallest allowable gross weight.

14) Continue with layout and structural design. Modify group weight statement as the design progresses.

15) Make detailed drag and vertical drag estimates based on drawings and model tests if possible.

16) Maintain close coordination between the team members to ensure that design decisions and design compromises are incorporated in the continual updating of the various related tasks.

17) The result.

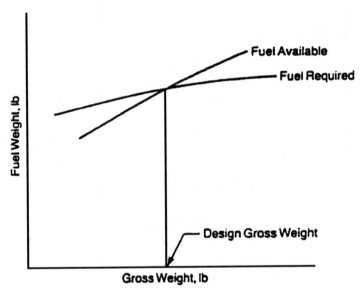

Result of preliminary design study. (Ref.: Prouty, R. W., *Military Helicopter Design Technology*, Jane's Defence Data, UK, 1989. Reproduced with permission from Jane's Information Group.)

The gross weight that makes the fuel available equal to the fuel required is the design gross weight (dGW) as shown. As a fallout of this process, the difference in the slopes of the two lines of fuel weight vs gross weight yields the growth factor (GF)—the change in gross weight that is forced by a one-pound increase in the payload or the structural weight. The growth factor is

$$GF = \frac{1}{\frac{dF_{avail}}{dGW} - \frac{dF_{req}}{dGW}}, \ lb/lb$$

where the two slopes are taken at the design gross weight. The denominator is always less than unity; so the growth factor is always greater than unity.

Helicopter Design, continued

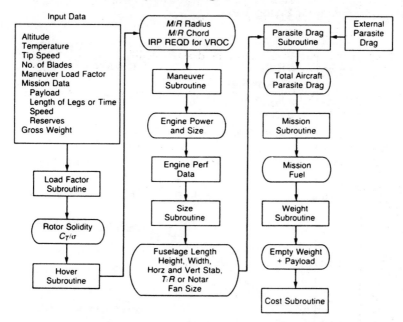

Block diagram of typical computer program for 13 initial steps of helicopter preliminary design. (Ref.: Prouty, R. W., *Military Helicopter Design Technology*, Jane's Defence Data, UK, 1989. Reproduced with permission from Jane's Information Group.)

Helicopter Design, continued

Tail Rotor Diameter Sizing Trend

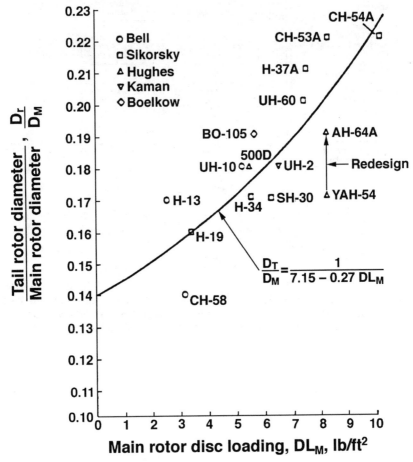

(Ref.: Wiesner and Kohler, *Tail Rotor Design Guide*, U.S. Army Air Mobility Research and Development Laboratory, Fort Eustis, VA, USAAMRDL TR 73-99, 1973.)

Helicopter Design, continued

Performance

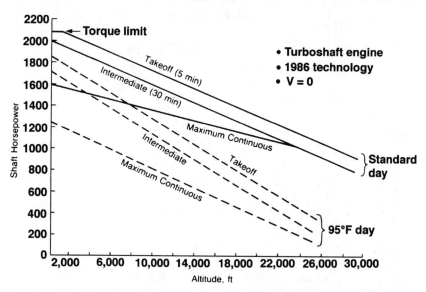

Typical uninstalled engine ratings as a function of altitude and temperature. (Ref.: Figure 4.1, page 274, from *Helicopter Performance, Stability, and Control*, by Raymond W. Prouty, 1995, Krieger Publishing Co., Malabar, Florida.)

Helicopter Design, continued

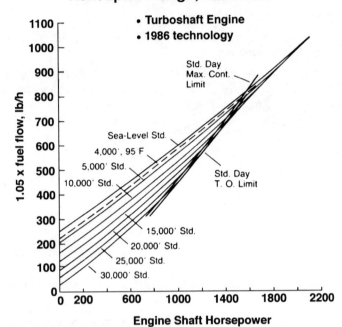

Typical engine fuel flow characteristics. (Ref.: Figure 4.3, page 276, from *Helicopter Performance, Stability, and Control*, by Raymond W. Prouty, 1995, Krieger Publishing Co., Malabar, Florida.)

Helicopter Design, continued

Basic Flight Loading Conditions

Critical flight loading conditions normally considered in the design of a pure helicopter are defined as follows.

- Maximum speed (design limit speed V_H)
- Symmetrical dive and pullout at design limit speed V_{DL} and at 0.6 V_H, approximately the speed of maximum load factor capability
- Vertical takeoff (jump takeoff)
- Rolling pullout
- Yaw (pedal kicks)
- Autorotational maneuvers

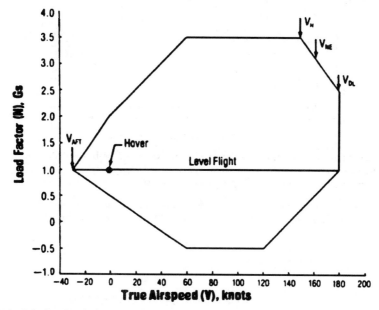

A typical design *V–N* diagram. (Ref.: Prouty, R. W., *Helicopter Aerodynamics* (*Rotor & Wing International*), PJS Publications, 1985. Reproduced with permission from Raymond W. Prouty.)

These, and other limits, are normally set by the customer or certifying authority and are depicted in the velocity-load (*V–N*) diagram. Other parameters usually defined in the *V–N* diagram are never-to-exceed velocity (V_{NE}) and maximum rearward velocity (V_{AFT}). The design structural envelope must satisfy the *V–N* diagram limits.

Helicopter Design, continued

Rotor Thrust Capabilities

The maximum rotor thrust capabilities are shown below.

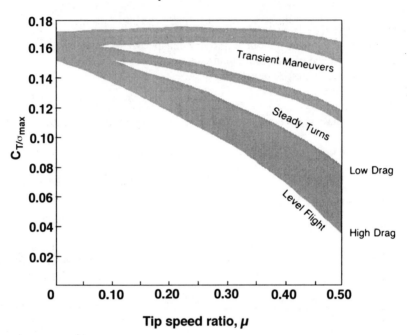

Rotor thrust capability. (Ref.: Figure 5.2, page 345, from *Helicopter Performance, Stability, and Control*, by Raymond W. Prouty, 1995, Krieger Publishing Co., Malabar, Florida.)

Rotor thrust capability

$$\frac{C_T}{\sigma} = \frac{\text{Rotor thrust}}{\text{Density of air} \times \text{blade area} \times (\text{tip speed})^2}$$

and tip speed ratio

$$\mu = \frac{\text{Forward speed of helicopter}}{\text{Tip speed}}$$

Helicopter Design, continued

One-Hour Helicopter Design Process

Requirements

As an illustration of the procedure, it is helpful to use a specific example. This will be a small battlefield transport helicopter designed to meet the following performance requirements.

- Payload: five fully-equipped troops @ 228 lb = 1760 lb
- Crew: two people @ 200 lb = 400 lb
- Maximum speed at sea level: 200 kn at the 30-min rating
- Cruise speed at sea level: at least 175 kn at the maximum continuous engine rating
- Range: 300 n mile at continuous engine rating with 30-min fuel reserve
- Vertical rate-of-climb: 450 ft/min @ 4000 ft, 95°F, with 95% of 30-min rating
- Engines: two with sea level maximum continuous rating of 650 hp each, 30-min rating of 800 hp each.

Initial Gross Weight Estimating

Estimate the fuel required to do the mission. Assume a specific fuel consumption of 0.5 lb per hp-h. For cruise at continuous engine rating and 175 kn for 300 n mile, this gives 440 lb including reserve. When added to the payload and the crew weight, the resultant "useful load" is 3600 lb.

Estimate the gross weight using the Historic Trend of Useful Load to Gross Weight curve. Two lines are shown on the curve, one for commercial helicopters and a slightly lower one for combat helicopters, which are penalized by the necessity to carry things such as redundant components, armor protection, and self-sealing fuel tanks. For a design of the 1990's, the ratio for the combat helicopter has been chosen as 0.5, which means that the example helicopter should have a gross weight of about 7200 lb.

Check on Maximum Forward Speed

Estimate the drag characteristics by using the statistical trend for equivalent plate area as shown in the Statistical Trend for Equivalent Flat Plate Area curve, which is based on a number of existing helicopters. Choosing to use the line labeled "Average Drag," the helicopter will have an equivalent flat plate area of 16 ft². The maximum speed can be estimated by assuming that 70% of the installed power is being used to overcome parasite drag at high speed using the following equation.

$$\text{Max Spd} = 41\sqrt{\frac{\text{30-min rating of both engines}}{\text{equivalent flat plate area}}}, \text{ knots}$$

Bibliography

Abbot, I. H., and Von Doenhoff, A. E., *Theory of Wing Sections*, Dover, New York, 1958.

AMCP 706-201, *Engineering Design Handbook, Helicopter Engineering, Part One: Preliminary Design*, Headquarters, U.S. Army Materiel Command, Aug. 1974.

Anon., "Jet Transport Performance Methods," The Boeing Co., Document D6-1460, 1967.

Bligh, J. A., "Methods Report for Aircraft Performance Estimation," Northrop Aircraft, Rept. MR-1, 1964.

Corning, G., *Supersonic and Subsonic, CTOL and VTOL, Airplane Design*, published by author, 1960.

Dommasch, D. O., *Airplane Aerodynamics*, Pitman, London, 1961.

Etkin, B., *Dynamics of Flight, Stability and Control*, Wiley, New York, 1959.

Hill, P. G., and Peterson, C. R., *Mechanics and Thermodynamics of Propulsion*, Addison–Wesley, Reading, MA, 1965.

Hoerner, S. F., *Fluid-Dynamic Drag*, published by author, 1965.

Jenn, D. C., *Radar and Laser Cross Section Engineering*, AIAA Education Series, AIAA, Washington, DC, 1995.

Nicolai, L. M., *Fundamentals of Aircraft Design*, METS, Inc., 1975.

Oates, G. C., *Aerothermodynamics of Gas Turbine and Rocket Propulsion*, AIAA Education Series, AIAA, New York, 1984.

Oates, G. C., Editor, *Aerothermodynamics of Aircraft Engine Components*, AIAA Education Series, AIAA, New York, 1985.

Perkins, C. D., and Hage, R. E., *Airplane Performance Stability and Control*, Wiley, New York, 1967.

Pope, A., *Aerodynamics of Supersonic Flight*, Pitman, London, 1958.

Prouty, R. W., *Helicopter Aerodynamics* (*Rotor & Wing International*), PJS Publications, Inc., 1985.

Prouty, R. W., *Helicopter Performance, Stability, and Control*, 1995 Edition, Krieger, Melbourne, FL, 1995.

Prouty, R. W., *Military Helicopter Design Technology*, Jane's Defence Data, UK, 1989.

Seckel, E., *Stability and Control of Airplanes and Helicopters*, Academic, New York, 1964.

Torenbeek, E., *Synthesis of Subsonic Airplane Design*, Delft Univ., Delft, Netherlands, 1981.

Wood, K. D., *Aerospace Vehicle Design—Volume I: Aircraft Design*, Johnson Publishing Co., 1966.

Section 8

GEOMETRIC DIMENSIONING AND TOLERANCING (ANSI/ASME Y14.5M-1994)

ANSI/ASME Y14.5-1994

The ANSI/ASME Y14.5, revised in 1994, is an accepted geometric dimensioning standard followed throughout the industry. It is based on a philosophy of establishing datums and measuring features following the same procedures one would use to inspect the physical part.

Abbreviations and Acronyms

The following abbreviations and acronyms are commonly used in the industry.

ANSI	= American National Standards Institute
ASA	= American Standards Association
ASME	= American Society of Mechanical Engineers
AVG	= average
CBORE	= counterbore
CDRILL	= counterdrill
CL	= center line
CSK	= countersink
FIM	= full indicator movement
FIR	= full indicator reading
GD&T	= geometric dimensioning and tolerancing
ISO	= International Standards Organization
LMC	= least material condition
MAX	= maximum
MDD	= master dimension definition
MDS	= master dimension surface
MIN	= minimum
mm	= millimeter
MMC	= maximum material condition
PORM	= plus or minus
R	= radius
REF	= reference
REQD	= required
RFS	= regardless of feature size
SEP REQT	= separate requirement
SI	= Système International (the metric system)
SR	= spherical radius
SURF	= surface
THRU	= through
TIR	= total indicator reading
TOL	= tolerance

ANSI/ASME Y14.5-1994, continued

Definitions

Datum—a theoretically exact point, axis, or plane derived from the true geometric counterpart of a specified datum feature. A datum is the origin from which the location and geometric characteristics or features of a part are established.

Datum feature—an actual feature of a part that is used to establish a datum.

Datum target—a specified point, line, or area on a part used to establish a datum.

Dimension, basic—a numerical value used to describe the theoretically exact size, profile, orientation, or location of a feature or datum target. It is the basis from which permissible variations are established by tolerances on other dimensions, in notes, or in feature control frames.

Feature—the general term applied to a physical portion of a part, such as a surface, pin, tab, hole, or slot.

Least material condition (LMC)—the condition in which a feature of size contains the least amount of material within the stated limits of size, e.g., maximum hole diameter, minimum shaft diameter.

Maximum material condition (MMC)—the condition in which a feature of size contains the maximum amount of material within the stated limits of size, e.g., minimum hole diameter, maximum shaft diameter.

Regardless of feature size (RFS)—the term used to indicate that a geometric tolerance or datum reference applies at any increment of size of the feature within its size tolerance.

Tolerance, geometric—the general term applied to the category of tolerances used to control form, profile, orientation, location, and runout.

True position—the theoretically exact location of a feature established by basic dimensions.

Geometric Symbols, Definitions, and Modifiers

Type of Tolerance	Geometric Symbol	Definition	Modifier Allowed	Datum Reference	Datum Modifiers Allowed
Form	—	Straightness	YES	no	N/A
Form	▱	Flatness	NO	no	N/A
Form	◯	Circularity	NO	no	N/A
Form	⌭	Cylindricity	NO	no	N/A
Profile	⌓	Profile / Surface	RFS	YES	YES
Profile	⌒	Profile / Line	RFS	YES	YES
Orientation	//	Parallelism	YES	YES	YES
Orientation	⊥	Perpendicularity	YES	YES	YES
Orientation	∠	Angularity	YES	YES	YES
Location	⊕	True Position	YES	YES	YES
Location	◎	Concentricity	Ⓢ	YES	Ⓢ
Location	⌖	Symmetry	Ⓢ	YES	Ⓢ
Runout	↗	Circular Runout	NO	YES	NO
Runout	↗↗	Total Runout	NO	YES	NO

Miscellaneous Definitions

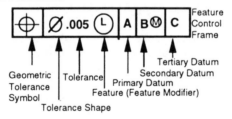

Feature Control Frame

Geometric Tolerance Symbol
Tolerance Shape
Tolerance
Feature (Feature Modifier)
Primary Datum
Secondary Datum
Tertiary Datum

⌜.500⌟ Basic (Exact Theoretical) Dimension

A numerical value used to describe the theoretically exact size, profile, orientation, or location of a feature or datum target. It is the basis from which permissible variations are established by tolerances on other dimensions, in notes, or in feature control frames.

Ø   **Cylindrical Tolerance Zone or Diameter Symbol**

A   **Datum Feature Symbol**

A datum is the origin from which the location or geometric characteristics of features of a part are established. Datums are theoretically exact points, axes, or planes derived from the true geometric counterpart of a specified datum feature.

Material Condition Symbols and Definitions

Ⓜ Maximum Material Condition (MMC)

Condition in which a feature of size contains the maximum amount of material allowed by the size tolerance of the feature. For example, minimum hole diameter or maximum shaft diameter.

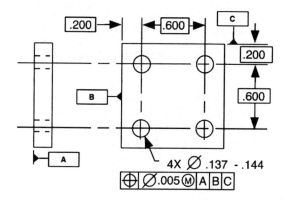

The Ⓜ in the feature control frame invokes the MMC concept and allows an increase in the amount of positional tolerance as the features depart from the maximum material condition. Note: This example, which is used to demonstrate the least material condition would hardly ever be seen in real practice. Replacement of the four holes with a shaft of the same diameter would be more typical.

Hole size	Tolerance zone
0.137	0.005
0.138	0.006
0.139	0.007
0.140	0.008
0.141	0.009
0.142	0.010
0.143	0.011
0.144	0.012

Material Condition Symbols and Definitions, continued

Ⓛ Least Material Condition (LMC)

Condition in which a feature of size contains the minimum amount of material allowed by the size tolerance of the feature. For example, maximum allowable hole diameter or minimum shaft diameter.

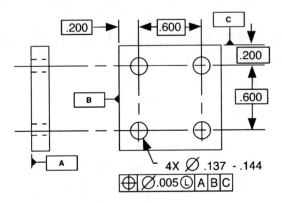

The Ⓛ in the feature control frame invokes the LMC concept and allows an increase in the amount of positional tolerance as the features depart from the least material condition. Note: If nothing appears in the feature control frame following the tolerance, then the RFS condition is assumed.

Hole size	Tolerance zone
0.144	0.005
0.143	0.006
0.142	0.007
0.141	0.008
0.140	0.009
0.139	0.010
0.138	0.011
0.137	0.012

Material Condition Symbols and Definitions, continued

Regardless of Feature Size (RFS)

The term used to indicate that a geometric tolerance or datum reference applies at any increment of size of the feature within its size tolerance.

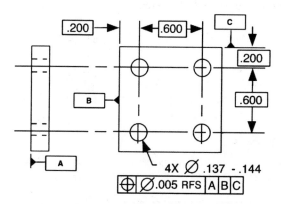

The RFS in the feature control frame invokes the RFS concept and allows no increase in the amount of positional tolerance as the features depart from the maximum material condition.

Hole size	Tolerance zone
0.137	0.005
0.138	0.005
0.139	0.005
0.140	0.005
0.141	0.005
0.142	0.005
0.143	0.005
0.144	0.005

Standard Rules

Limits of Size Rule

Where only a tolerance of size is specified, the limits of size of the individual feature prescribe the extent to which variations in its geometric form as well as size are allowed.

Tolerance Rule

For all applicable geometric tolerances, RFS applies with respect to the individual tolerance, datum reference, or both, where no modifying symbol is specified. Ⓜ or Ⓛ must be specified on the drawing where it is desired, when applicable.

Pitch Diameter Rule

Each tolerance of orientation or position and datum reference specified for a screw thread applies to the axis of the thread derived from the pitch diameter.

(MEANS; APPLIES TO THE PITCH DIAMETER)

Datum/Virtual Condition Rule

Depending on whether it is used as a primary, secondary, or tertiary datum, a virtual condition exists for a datum feature of size where its axis or centerplane is controlled by a geometric tolerance and referenced at MMC. In such a case, the datum feature applies at its virtual condition even though it is referenced in a feature control frame at MMC.

Standard Rules, continued

Additional Symbols

REFERENCE DIMENSION	(50)
PROJECTED TOLERANCE ZONE	Ⓟ
DATUM TARGET	Ⓐ̄₁
DATUM TARGET POINT	✕
DIMENSION ORIGIN	⊕▶
CONICAL TAPER	◁—
SLOPE	◸
COUNTERBORE/SPOTFACE	⊔
COUNTERSINK	∨
DEPTH/DEEP	↧
SQUARE (SHAPE)	□
DIMENSION NOT TO SCALE	<u>15</u>
NUMBER OF TIMES OR PLACES	8X
ARC LENGTH	⌒105
RADIUS	R
SPHERICAL RADIUS	SR
SPHERICAL DIAMETER	S⌀
ALL AROUND (PROFILE)	⌀—
BETWEEN SYMBOL	◀—▶
TANGENT PLANE	Ⓣ
STATISTICALLY DERIVED VALUE	⟨ST⟩

⊕ Location Tolerance

Drawing Callout

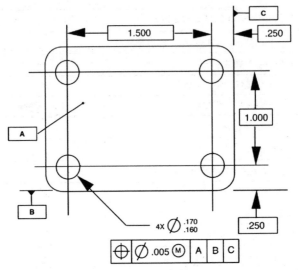

Interpretation

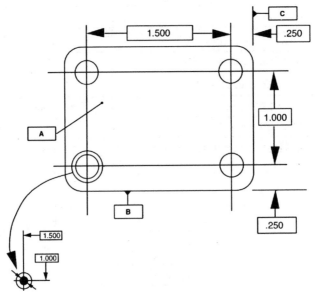

⊕ Location Tolerance, continued

Basic dimensions establish the true position from specified datum features and between interrelated features. A positional tolerance defines a zone in which the center, axis, or centerplane of a feature of size is permitted to vary from the true position.

- If hole size is 0.160, then the axis of hole must lie within 0.005 dia tol zone.
- If hole size is 0.170, then the axis of hole must lie within 0.015 dia tol zone.

Mating Parts—Fixed Fastener

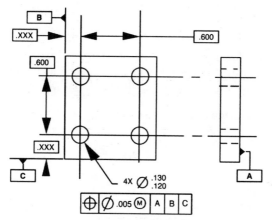

Part 1—Clearance holes.

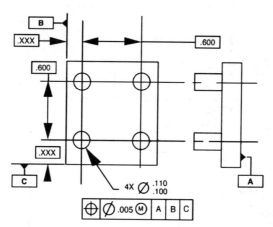

Part 2—Fixed studs. (Inserts, pins, fixed nut plates, countersunk holes.)

⊕ Location Tolerance, continued

Calculations

To determine positional tolerance if hole and stud size are known and projected tolerance zone is used:

Positional tolerance calculation—fixed fastener
$$T = \frac{H-S}{2}$$

Hole MMC (H)   0.120
Stud MMC (S) $(-)$ $\dfrac{0.110}{}$
$$\frac{0.010}{2} = 0.005$$

0.005 position tolerance on all holes and studs (or any combination on each part that totals 0.010)

When projected tolerance zone is used,

$$T_1 + T_2\left(1 + \frac{2P}{D}\right) = H - S$$

where

T_1 = positional tolerance diameter of hole
T_2 = positional tolerance diameter of tapped hole
D = minimum depth of engagement
P = maximum projection of fastener
S = stud diameter (MMC)
H = hole diameter (MMC)

Mating Parts—Floating Fastener

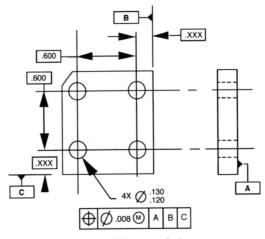

Part 1—Clearance holes.

⊕ Location Tolerance, continued

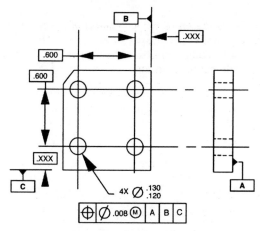

Part 2—Clearance holes.

Calculations

To determine positional tolerance if hole and fastener size are known:

Positional tolerance calculation—floating fastener	
$T = H - F$	
Hole MMC	0.120
Fastener MMC	(−) 0.112
Positional tolerance at MMC —both parts	0.008

—— Straightness of a Surface

Definition

Straightness is the condition where an element of a surface or axis is a straight line. A straightness tolerance specifies a tolerance zone within which the considered element or axis must lie and is applied in the view where the elements to be controlled are represented by a straight line.

Tol type	Symbol	Datum REF	Implied cond.	Allowable modifiers	Tol zone shape
Form	——	None	N/A for surfaces RFS Rule 3 for all axes or centerplane	ⓜ if applied to an axis or centerplane	Parallel lines (surface) Parallel planes (centerplane) Cylindrical (axis)

Comment

. Is additive to size when applied to an axis.

Drawing Callout

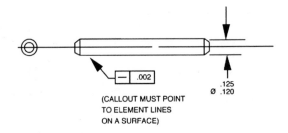

(CALLOUT MUST POINT
TO ELEMENT LINES
ON A SURFACE)

.125
Ø .120

— .002

——— Straightness of a Surface, continued

Interpretation

All elements of the surface must lie within a tolerance zone defined by two perfectly straight parallel lines 0.002 apart. Additionally, the part must be within the perfect form boundary (Limits of Size Rule).

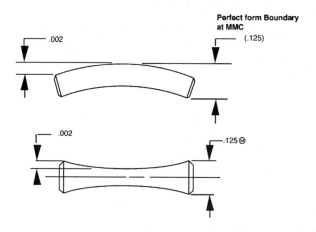

——— Straightness of an Axis

Definition

Straightness is the condition where an element of a surface or axis is a straight line. A straightness tolerance specifies a tolerance zone within which the considered element or axis must lie and is applied in the view where the elements to be controlled are represented by a straight line.

Tol type	Symbol	Datum REF	Implied cond.	Allowable modifiers	Tol zone shape
Form	———	None	N/A for surfaces RFS Rule 3 for all axes or centerplane	Ⓜ if applied to an axis or centerplane	Parallel lines (surface) Parallel planes (centerplane) Cylindrical (axis)

Comment

• Is additive to size when applied to an axis.

___ Straightness of an Axis, continued

Drawing Callout

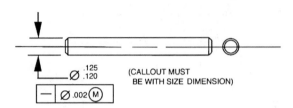

$\varnothing$ .125 / .120

(CALLOUT MUST BE WITH SIZE DIMENSION)

| — | $\varnothing$ .002 Ⓜ |

Interpretation

The axis of the feature must be contained by a cylindrical tolerance zone of 0.002 diameter when the pin is at MMC.

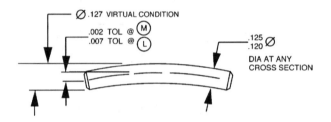

$\varnothing$ .127 VIRTUAL CONDITION

.002 TOL @ Ⓜ
.007 TOL @ Ⓛ

.125 / .120 $\varnothing$

DIA AT ANY CROSS SECTION

▱ Flatness

Definition

A condition of a surface having all elements in one plane.

Tol type	Symbol	Datum REF	Implied cond.	Allowable modifiers	Tol zone shape
Form	▱	Never	N/A	N/A	Parallel planes

Comments

- No particular orientation.
- Not additive to size or location limits.

▱ Flatness, continued

Drawing Callout

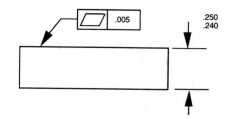

Interpretation

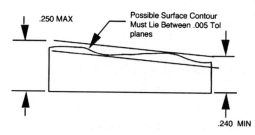

○ Circularity (Roundness)

Definition

A condition on a surface or revolution where all points of the surface intersected by any plane perpendicular to a common axis or center (sphere) are equidistant from the axis or center.

Tol type	Symbol	Datum REF	Implied cond.	Allowable modifiers	Tol zone shape
Form	○	Never	N/A	N/A	Conc. circles

Comment

- Applies at single cross sections only.

O Circularity (Roundness), continued

Drawing Callout

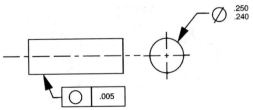

Interpretation

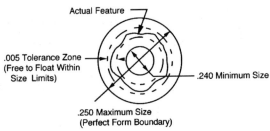

Actual Feature

.005 Tolerance Zone
(Free to Float Within
Size Limits)

.240 Minimum Size

.250 Maximum Size
(Perfect Form Boundary)

⌭ Cylindricity

Definition

A condition on a surface of revolution in which all points of the surface are equidistant from a common axis.

Tol type	Symbol	Datum REF	Implied cond.	Allowable modifiers	Tol zone shape
Form	⌭	Never	N/A	N/A	Conc. cylnds.

Comment

- Applies over entire surface.

⌀ Cylindricity, continued

Drawing Callout

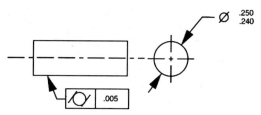

Interpretation

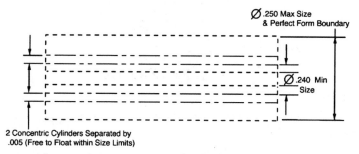

2 Concentric Cylinders Separated by
.005 (Free to Float within Size Limits)

// Parallelism

Definition

The condition of a surface or axis which is equidistant at all points from a datum plane or datum axis.

Tol type	Symbol	Datum REF	Implied cond.	Allowable modifiers	Tol zone shape
Orientation	//	Always	Ⓢ RFS	Ⓜ or Ⓛ if feature has a size consideration	Parallel planes (surface) Cylindrical (axis)

Comment

- Parallelism tolerance is not additive to feature size.

// Parallelism, continued

Drawing Callout

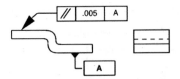

Interpretation

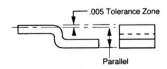

⌒ Profile of a Line

Definition

Specifies a uniform boundary along the true profile within which the elements of the surface must lie.

Tol type	Symbol	Datum REF	Implied cond.	Allowable modifiers	Tol zone shape
Profile	⌒	W/WO datum	RFS	Ⓜ or Ⓛ may be applied to the datum of REF only	Two lines disposed about the theoretical exact profile

Comments

- Always relates to a theoretically exact profile.
- Profile is the only characteristic that can be used with or without a datum of reference.
- Two-dimensional tolerance zone.
- Applies only in the view in which the element is shown as a line.

⌒ Profile of a Line, continued

Drawing Callout

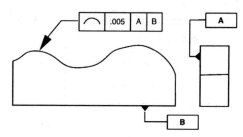

Interpretation

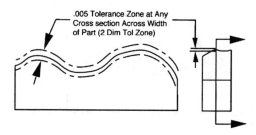

.005 Tolerance Zone at Any
Cross section Across Width
of Part (2 Dim Tol Zone)

⌓ Profile of a Surface

Definition

Specifies a uniform boundary along the true profile within which the elements of the surface must lie.

Tol type	Symbol	Datum REF	Implied cond.	Allowable modifiers	Tol zone shape
Profile	⌓	W/WO datum	RFS	Ⓜ or Ⓛ may be applied to the datum of REF only	Two surfaces disposed about the theoretical exact contour

Comments

- Always relates to a theoretically exact profile.
- Profile is the only characteristic that can be used with or without a datum of reference.
- Three-dimensional tolerance zone.
- Applies to entire surface shown.

⌓ Profile of a Surface, continued

Drawing Callout

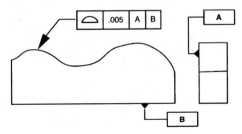

Interpretation

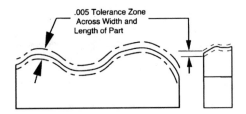

⊥ Perpendicularity

Definition

Condition of a surface, axis, median plane, or line which is exactly at 90 deg with respect to a datum plane or axis.

Tol type	Symbol	Datum REF	Implied cond.	Allowable modifiers	Tol zone shape
Orientation	⊥	Always	RFS	Ⓜ or Ⓛ if feature has a size consideration	Parallel plane (surface or centerplane) Cylindrical (axis)

Comment

- Relation to more than one datum feature should be considered to stabilize the tolerance zone in more than one direction.

⊥ Perpendicularity, continued

Drawing Callout

Example 1

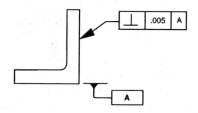

Interpretation

Example 1

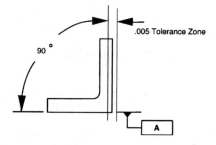

Drawing Callout

Example 2

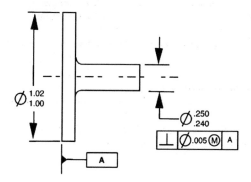

⊥ Perpendicularity, continued

Interpretation

Example 2

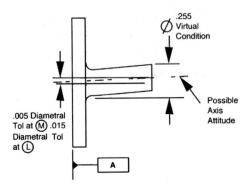

.255
∅ Virtual
Condition

Possible
Axis
Attitude

.005 Diametral
Tol at Ⓜ .015
Diametral Tol
at Ⓛ

A

∠ Angularity

Definition

Condition of a surface, or axis, at a specified angle, other than 90 deg from a datum plane or axis.

Tol type	Symbol	Datum REF	Implied cond.	Allowable modifiers	Tol zone shape
Orientation	∠	Always	RFS	Ⓜ or Ⓛ if feature has a size consideration	Parallel plane (surface) Cylindrical (axis)

Comments

- Always relates to basic angle.
- Relation to more than one datum feature should be considered to stabilize the tolerance zone in more than one direction.

∠ Angularity, continued

Drawing Callout

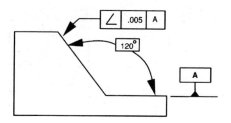

Interpretation

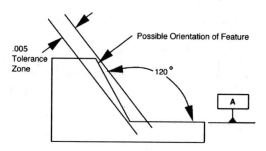

⊚ Concentricity

Definition

Condition where the axes at all cross-sectional elements of a surface of revolution are common to a datum axis.

Tol type	Symbol	Datum REF	Implied cond.	Allowable modifiers	Tol zone shape
Location	⊚	Always	RFS	N/A	Cylindrical

Comment

- Must compare axes, very expensive, first try to use position or runout.

⊚ Concentricity, continued

Drawing Callout

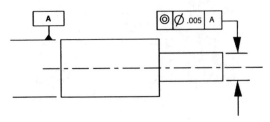

Interpretation

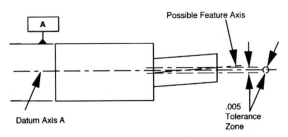

═ Symmetry

Definition

Condition where median points of all opposed elements of two or more feature surfaces are congruent with the axis or center plane of a datum feature.

Tol type	Symbol	Datum REF	Implied cond.	Allowable modifiers	Tol zone shape
Location	═	Always	RFS	N/A	Parallel plane (surface)

Comment

- May only be specified on an RFS basis.

Symmetry, continued

Drawing Callout

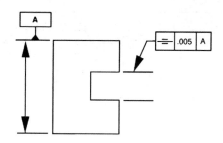

Interpretation

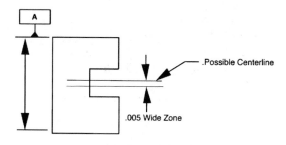

⤸ Circular Runout

Definition

A composite tolerance used to control the relationship of one or more features of a part to a datum axis during a full 360-deg rotation.

Tol type	Symbol	Datum REF	Implied cond.	Allowable modifiers	Tol zone shape
Runout	⤸	Yes	RFS	None	Individual circular elements must lie within the zone

Comments

- Simultaneously detects the combined variations of circularity and coaxial misregistration about a datum axis.
- FIM defined as full indicator movement.

⚹ Circular Runout, continued

Drawing Callout

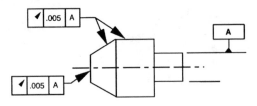

Interpretation

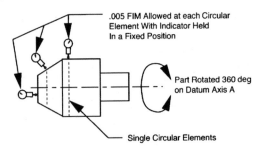

⚹⚹ Total Runout

Definition

A composite tolerance used to control the relationship of several features at once, relative to a datum axis.

Tol type	Symbol	Datum REF	Implied cond.	Allowable modifiers	Tol zone shape
Runout	⚹⚹	Always	RFS	N/A	⚹⚹ .005 A FIM

Comments

- Can be defined as the relationship between two features.
- May only be specified on an RFS basis.

Simultaneously detects combined errors of circularity, cylindricity, straightness, taper, and position.

⚿ Total Runout, continued

Drawing Callout

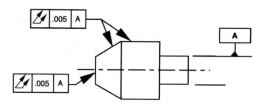

Interpretation

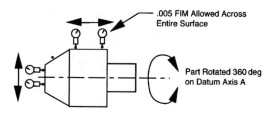

.005 FIM Allowed Across
Entire Surface

Part Rotated 360 deg
on Datum Axis A

⊕ Positioning for Symmetry

Definition

A condition in which a feature is symmetrically disposed about the centerline
of a datum feature.

Tol type	Symbol	Datum REF	Implied cond.	Allowable modifiers	Tol zone shape
Location	⊕	Yes	None	Ⓜ or Ⓛ	Two parallel planes

⊕ Positioning for Symmetry, continued

Drawing Callout

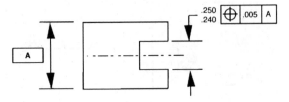

Interpretation

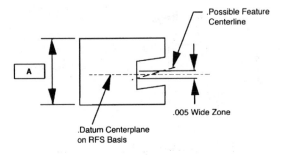

Composite Positional Tolerancing

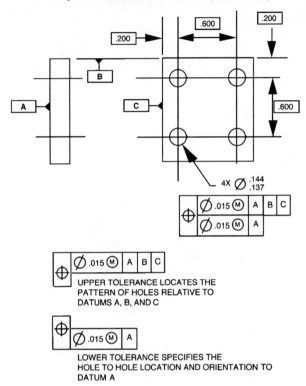

UPPER TOLERANCE LOCATES THE
PATTERN OF HOLES RELATIVE TO
DATUMS A, B, AND C

LOWER TOLERANCE SPECIFIES THE
HOLE TO HOLE LOCATION AND ORIENTATION TO
DATUM A

⊕ Datum Targets

Definition

A specified point, line, or area used to establish a datum, plane, or axis.

Datum Target Point

A datum target point is indicated by the symbol X which is dimensionally located using the other datums that compose the datum reference frame on a direct view of the surface.

⊖ Datum Targets, continued

Drawing Callout

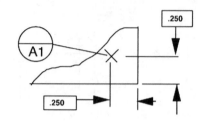

Interpretation

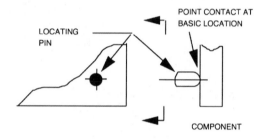

Datum Target Line

A datum target line is indicated by the symbol X on an edge view of the surface and a phantom line on the direct view or both.

Drawing Callout

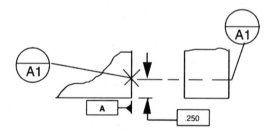

⊕ Datum Targets, continued

Interpretation

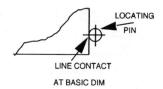

Datum Target Area

A datum target area is indicated by section lines inside a phantom outline of the desired shape with controlling dimensions added. The datum target area diameter is given in the upper half of the datum target symbol.

Drawing Callout

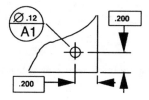

Interpretation

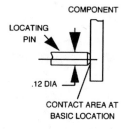

When datum targets are located using basic dimensions, standard gage or tool tolerances apply.

⊖ Example Datum Targets

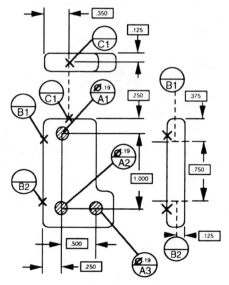

Typical drawing callout using datum targets to establish a datum reference frame for all three axes.

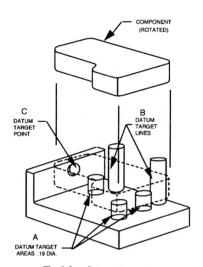

Tool for datum targets.

Manufacturing Cost vs Tolerance

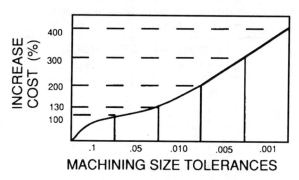

MACHINING SIZE TOLERANCES

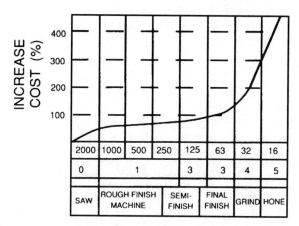

MACHINING FINISH TOLERANCES

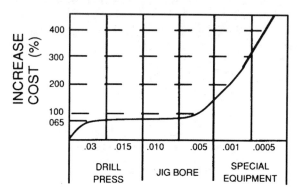

HOLE LOCATION TOLERANCES

Conversion Chart
Diametral True Position Tolerance to Coordinate Tolerance Transform

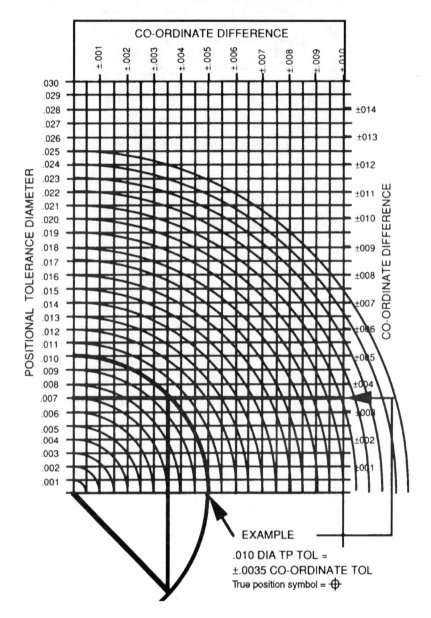

EXAMPLE

.010 DIA TP TOL =
±.0035 CO-ORDINATE TOL
True position symbol = ⊕

Conversion of True Position Tolerance Zone to/from Coordinate Tolerance Zone

Conversion Formula

$$2\sqrt{X^2 + Y^2} = \oplus$$

where

X, Y = coordinate tolerance ($\pm$)

$\oplus$ = diametral true position tolerance

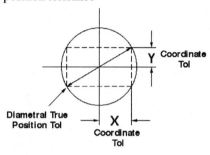

This conversion may only be used if the designer takes into account the tolerance across corners of a square zone. If not, a conversion of this type could result in a round tolerance zone much larger than is really allowed.

Section 9

MATERIALS AND SPECIFICATIONS

Metallic Materials

Material	F_{tu},[a] ksi	F_{ty},[b] ksi	F_{cy},[c] ksi	E_t,[d] psi/10^6	w,[e] lb/in.^{3}	Characteristics
Alloy steels						
4130 normalized	95	75	75	29	0.283	≤0.187 thick ⎫
sheet, strip, plate,						⎬ weldable
and tubing	90	70	70	29	0.283	>0.187 thick ⎭
4130 wrought forms						
(180 H.T.)	180	163	173	29	0.283	<0.5 equiv. diam, weldable
4330 wrought forms						
(220 H.T.)	220	186	194	29	0.283	<2.5 equiv. diam
DGAC wrought forms						
(220 H.T.)	220	190	198	29	0.283	<5.0 equiv. diam
300 M bars, forgings,						
and tubing (280 H.T.)	280	230	247	29	0.283	<5.0 equiv. diam
4340 bars, forgings,						
and tubing (260 H.T.)	260	215	240	29	0.283	<3.5 equiv. diam
Stainless steels						
301 (full hard) sheet						Weldable
and strip	185	140	98*	26	0.286	*Longitudinal grain direction
15-5 PH bars	115	75	99			Readily forged and welded
and forgings	190	170	143	28.5	0.283	Available in range of H.T.
PH15-7 Mo sheet,						
strip, and plate	190	170	179	29	0.277 ⎫	Readily cold formed and cold
bars and forgings	180	160	168	29	0.277 ⎭	drawn
17-4 PH sheet, strip,	135	105			0.284 ⎫	
and plate	190	170		28.5	0.282	Readily forged, welded, and
bars	195	75			0.284	brazed
	190	170		28.5	0.282	Can be sand or investment mold-
castings	130	120				ed or centrifugally cast
	180	160		28.5	0.282 ⎭	
Heat resistant steels						
A286 sheet, strip,						
and plate	140	95	95	29.1	0.287 ⎫	High strength up to 1300°F
bars, forgings, tubing	130	85	85	29.1	0.287 ⎭	Weldable
Inconel 600						
sheet, strip, plate,						
tubing, and forgings	80	30	30	30	0.304	Annealed ⎫ For low stres-
						⎬ sed parts up to
	95	70				2000°F—weld-
Bars and rods	120	90		30	0.304	Cold drawn ⎭ able
Inconel 718						
sheet, plate, and						High strength and creep resistant
tubing	170	145		29.6	0.297	to 1300°F—can be cast (lower
bars and forgings	180	150		29.6	0.297	values)

(continued)

[a] F_{tu} = strength, tensile ultimate. [b] F_{ty} = strength, tensile yield. [c] F_{cy} = strength, compressive yield. [d] E_t = tangent modulus. [e] w = density.

Source: MIL-HDBK-5, "Metallic Materials and Elements for Aerospace Vehicle Structures."

Metallic Materials, continued

Material	F_{tu},[a] ksi	F_{ty},[b] ksi	F_{cy},[c] ksi	E_t,[d] psi/10^6	w,[e] lb/in.^{3}	Characteristics
Aluminum alloy sheet						
2024-T3 (bare)	63	42	45	10.7	0.100	Common use—low cost
2024-T3 (clad)	58	39	42	10.5	0.100	Good strength/weight
2219-T87	62	50	50	10.5	0.102	High strength—creep resistant
5456-H343	53	41	39	10.2	0.096	Corrosion resistant—good weldability
6061-T6	42	36	35	9.9	0.098	Low cost—formable—weldable
7075-T6	76	66	67	10.3	0.101	High strength/weight
7075-T73	67	56	55	10.3	0.101	Stress, corrosion resistant
7178-T6	83	73	73	10.3	0.102	High strength/weight
Plate						
2024-T351	57	41	36	10.7	0.100	Common use—low cost
2219-T87	62	50	50	10.5	0.102	High strength—creep resistant
5456-H343	53	41	39	10.2	0.096	Corrosion resistant—good weldability
6061-T651	42	36	35	9.9	0.098	Low cost—formable—weldable
7050-T73651	71	62	60	10.3	0.102	Good fracture toughness
Extrusions						
2024-T4	60	44	39	10.7	0.100	Common use—low cost
6061-T6	38	35	34	10.1	0.098	Low cost—corrosion resistant—weldable
7050-T6510/1	68	59	64	10.3	0.102	High stress/corrosion resistance
7075-T6	81	73	74	10.5	0.101	High strength/weight
7075-T73	66	58	58	10.5	0.101	Good stress/corrosion resistance
7178-T6	88	79	79	10.5	0.102	High strength/weight
Tubing						
2024-T3	64	42	42	10.5	0.100	Low cost—common use
6061-T6	42	35	34	10.1	0.098	Weldable—corrosion resistant
Forgings						
2014-T6	65	55	55	10.7	0.101	Common use
7050-T736	70	54	57	10.2	0.102	Good fracture toughness
7075-T73	61	52	54	10.0	0.101	Good stress/corrosion resistance
Castings						
356-T6	30	20	20	10.4	0.097	Easy sand and investment
A356-T61	38	28	28	10.4	0.097	Good corrosion resistance
A357-T61	50	40	40	10.4	0.097	Premium castings
Titanium alloy						
6AL-4V (S.T.A.)						
sheet, strip, and plate	160	145	150	16.0	0.160	Can be spot and fusion welded
forgings (aww)	130	120		16.0	0.160	Corrosion resistant
bars	145	135		16.0	0.160	High strength
6AL-6V-2SN						
sheet, strip, and plate	170	160	170	17.0	0.164	High strength
forgings	150	140		17.0	0.164	Good formability
bars	170	155		17.0	0.164	Corrosion resistant

[a] F_{tu} = strength, tensile ultimate. [b] F_{ty} = strength, tensile yield. [c] F_{cy} = strength, compressive yield. [d] E_t = tangent modulus. [e] w = density.

Source: MIL-HDBK-5, "Metallic Materials and Elements for Aerospace Vehicle Structures."

Weights of Liquids

Liquid	Specific gravity at °C		Specific wt., lb/U.S. gal	lb/ft³
Alcohol (methyl)	0.810	0	6.75	50.5
Benzine	0.899	0	7.5	56.1
Carbon tetrachloride	1.595	20	13.32	99.6
Ethylene glycol	1.12		9.3	69.6
Gasoline	0.72		5.87	44.9
Glycerine	1.261	20	10.52	78.71
JP1	0.80		6.65	49.7
JP3	0.775		6.45	48.2
JP4	0.785		6.55	49.0
JP5	0.817	15	6.82	51.1
JP6	0.810		6.759	50.5
Kerosene	0.82		6.7	51.2
Mercury	13.546	20	113.0	845.6
Oil	0.89	15	7.4	55.3
Sea water	1.025	15	8.55	63.99
Synthetic oil	0.928	15	7.74	57.9
Water	1.000	4	8.345	62.43

Weights of Gases

Gas	Specific wt.,[a] lb/ft³	
Air	0.07651	(at 59.0°F)
Air	0.08071	
Carbon dioxide	0.12341	
Carbon monoxide	0.07806	
Helium	0.01114	
Hydrogen	0.005611	
Nitrogen	0.07807	
Oxygen	0.089212	

[a] At atmospheric pressure and 0°C.

Composite vs Metal Alloy[a]

Style of material Cure class	Maximum service temp, °F	Density, lb/in.³	Ultimate tensile strength, ksi	Tensile modulus, ×10⁶	Ultimate flexural strength, ksi	Flexural modulus of elasticity, psi ×10⁶	Specific tensile modulus, in. ×10⁶	Ultimate compression strength, ksi	Cured ply thickness, mil	Coefficient of thermal expansion, in./in./°F ×10⁶	Thermal conductivity, Btu/ft²/°F/h/ft
Aluminum 7075-T6	300+	0.100	83	10	83	10	100	83	—	13.5	70
Carbon steel 4310	600	0.289	190	30	190	30	104	190	—	6.3	22
Stainless steel 316	1000+	0.290	84	28	84	28	96.5	84	—	8.9	9.4
Titanium 6AL 4VA	1000	0.160	160	16.5	160	16.5	103	160	—	5.3	4.2
Magnesium HK 31 A-H2A	200	0.65	29	6.4	29	6.4	98.5	29	—	14.0	66
Lo temp epoxy fiberglass 250°F cure 7781 cloth	200	0.066	63	3.4	78	3.2	51.5	61	9–10	2.7	0.26
Hi temp epoxy fiberglass 350°F cure 7781 cloth	350	0.066	70	4.5	88	3.5	68	71	9–10	2.8	0.25
Expoxy/fiberglass UD tape 350°F cure	350	0.066	160	7.0	184	6.9	106.0	88	5	8.3	0.2
Phenolic/fiberglass 7781 cloth 350°F cure	350	0.066	44	3.1	66	3.5	47.5	45	9–10	6.0	0.18

(continued)

[a]The reported property values are nominal laminate averages and can change due to a number of variables. The data are offered for general information convenience only, subject to your verification.

Composite vs Metal Alloy,[a] continued

Style of material Cure class	Maximum service temp, °F	Density, lb/in.³	Ultimate tensile strength, ksi	Tensile modulus, ×10⁶	Ultimate flexural strength, ksi	Flexural modulus of elasticity, psi ×10⁶	Specific tensile modulus, in. ×10⁶	Ultimate compression strength, ksi	Cured ply thickness, mil	Coefficient of thermal expansion, in./in./°F ×10⁶	Thermal conductivity, Btu/ft²/°F/h/ft
Polyimide/fiberglass 7781 cloth condensation type	600	0.067	58	4.0	78	3.5	60.0	58	9–10	6.0	0.18
Polyimide/fiberglass 7781 cloth PMR-15 600°F cure	600	0.072	63	3.8	84	4.0	53.0	70	9–10	5.0	0.125
Epoxy/Kevlar cloth 285 style Kevlar 49 350°F cure	350	0.052	72	3.5	58	3.3	67.0	28	9–10	0.001	0.52
Epoxy/graphite cloth 350°F cure HM woven graphite	350	0.058	80	10.0	108	9.0	172	85	8–9	0.008	28
Epoxy/graphite HM UD tape 350°F cure	350	0.057	230	19.0	254	17.1	333.0	190	5	±0	40
BMI/graphite HM cloth	450	0.055	94	11.0	124	8.5	200.0	106	8–9	0.01	30.0
BMI/graphite HM UD tape	450	0.055	251	21.5	288	18.4	391.0	238	5	±0	42.0
Polyimide graphite HM cloth PMR-15	600	0.058	106	9.5	122	12.0	164.0	104	8	0.01	30

[a]The reported property values are nominal laminate averages and can change due to a number of variables. The data are offered for general information convenience only, subject to your verification.

Galvanic Series

Galvanic Series of Some Commercial Metals and Alloys in Seawater

↑ Noble or cathodic	Platinum
	Gold
	Graphite
	Titanium
	Silver
	Hastelloy C (62 Ni, 17 Cr, 15 Mo)
	18-8 stainless steel (passive)
	Chromium stainless steel 11–30% Cr (passive)
	Inconel (passive)
	Nickel (passive)
	Silver solder
	Monel
	Cupronickels (60–90 Cu, 40–10 Ni)
	Bronzes (Cu-Sn)
	Copper
	Brasses (Cu-Zn)
	Hastelloy B
	Inconel (active)
	Nickel (active)
	Tin
	Lead
	Lead–tin solders
	18-8 stainless steel (active)
	Ni-Resist (high Ni cast iron)
	Chromium stainless steel, 13% Cr (active)
	Cast iron
	Steel or iron
	2024 aluminum
Active or anodic	Cadmium
	Commercially pure aluminum
↓	Zinc
	Magnesium and magnesium alloys

Section 10

SPACECRAFT DESIGN

Spacecraft General Information

Space Missions

Characteristic	Relevant missions
Global perspective	Communications Navigation Weather Surveillance
Above the atmosphere	Scientific observations at all wavelengths
Gravity-free environment	Materials processing in space
Abundant resources	Space industrialization Asteroid exploration Solar power satellites
Exploration of space itself	Exploration of moon and planets Scientific probes Asteroid and comet missions

Ref.: J. Wertz, *Space Mission Engineering* (*Mission Design*), Microcosm, Inc., Torrance, CA, 1990.

Typical Spacecraft Program Sequence

- Mission requirements definition
- Conceptual design
- Preliminary design (pdr)
- Detailed design (cdr)
- Subsystem fabrication
- Integration and test (frr)
- Launch vehicle integration
- Launch
- Orbital verification
- Operational use

Spacecraft General Information, continued

Spacecraft Block Diagram

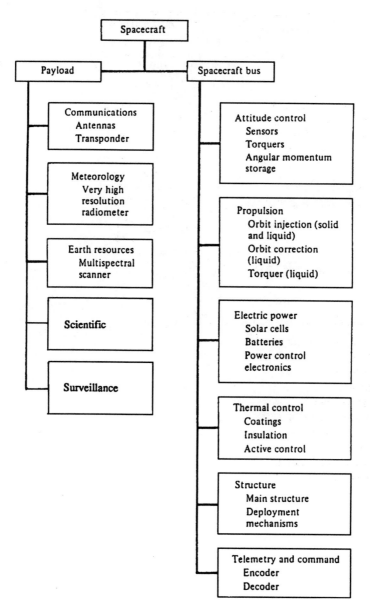

Spacecraft General Information, continued

Typical Spacecraft

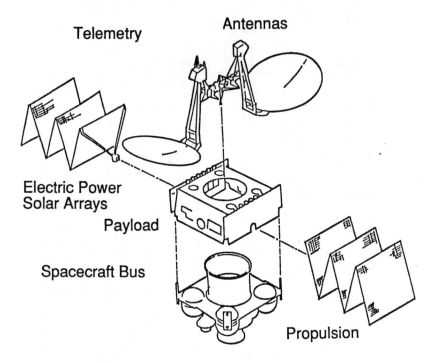

Launch Vehicles

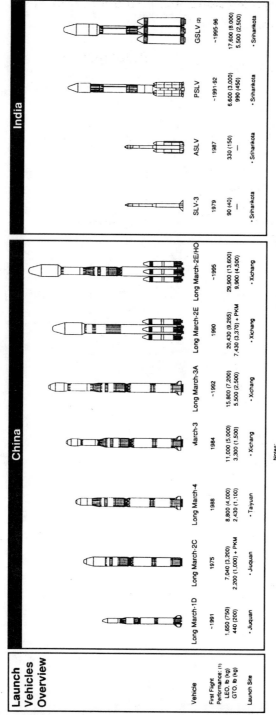

Launch Vehicles Overview

China

Vehicle	Long March-1D	Long March-2C	Long March-4	March-3	Long March-3A	Long March-2E	Long March-2E/HO
First Flight	~1991	1975	1988	1984	~1992	1990	~1995
Performance: (1) LEO, lb (kg)	1,650 (750)	7,040 (3,200)	8,800 (4,000)	11,000 (5,000)	15,800 (7,200)	20,430 (9,265)	29,900 (13,600)
GTO, lb (kg)	440 (200)	2,200 (1,000) + PKM	2,430 (1,100)	3,300 (1,500)	5,500 (2,500)	7,430 (3,370) + PKM	9,900 (4,500)
Launch Site	• Jiuquan	• Jiuquan	• Taiyuan	• Xichang	• Xichang	• Xichang	• Xichang

India

Vehicle	SLV-3	ASLV	PSLV	GSLV (2)
First Flight	1979	1987	~1991-92	~1995-96
Performance: (1) LEO, lb (kg)	90 (40)	330 (150)	6,600 (3,000)	17,600 (8,000)
GTO, lb (kg)	—	—	990 (450)	5,500 (2,500)
Launch Site	• Sriharikota	• Sriharikota	• Sriharikota	• Sriharikota

Notes:
(1) LEO - low Earth orbit
GTO - geosynchronous transfer orbit
GEO - geosynchronous orbit
PKM - perigee kick motor
(2) Configuration is being finalized
(3) LEO version is designated 6920 or 7920
(4) Plus 95% Polar with 10 solid strap-ons

(5) Plus 25% LEO with SRMU
(6) With Centaur upper stage
(7) Plus 10-20% LEO with ASRM
(8) Plus 15% LEO from Cape York
(9) Plus 35% GTO from Cape York
(10) With RCS kick stage
(11) With EUS upper stage

Launch Vehicles, continued

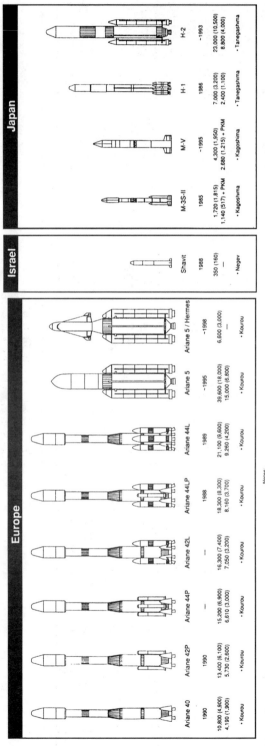

Japan

	M-3S-II	M-V	H-1	H-2
	1985	~1995	1986	~1993
	1,720 (1,815)	4,300 (1,950)	7,000 (3,200)	23,000 (10,500)
	1,140 (517) + PKM	2,680 (1,215) + PKM	2,400 (1,100)	8,800 (4,000)
	• Kagoshima	• Kagoshima	• Tanegashima	• Tanegashima

Israel

	Shavit
	1988
	350 (160)
	—
	• Negev

Europe

	Ariane 40	Ariane 42P	Ariane 44P	Ariane 42L	Ariane 44LP	Ariane 44L	Ariane 5	Ariane 5 / Hermes
	1990	1990	—	—	1988	1989	~1995	~1998
	10,800 (4,900)	13,400 (6,100)	15,200 (6,900)	16,300 (7,400)	18,300 (8,300)	21,100 (9,600)	39,600 (18,000)	6,600 (3,000)
	4,190 (1,900)	5,730 (2,600)	6,610 (3,000)	7,050 (3,200)	8,160 (3,700)	9,260 (4,200)	15,000 (6,800)	—
	• Kourou	• Kourou	• Kourou	• Kourou	• Kourou	• Kourou	• Kourou	• Kourou

Notes:
(1) LEO - low Earth orbit
GTO - geosynchronous transfer orbit
GEO - geosynchronous orbit
PKM - perigee kick motor
(2) Configuration is being finalized
(3) LEO version is designated 6920 or 7920
(4) Plus 95% Polar with 10 solid strap-ons

(5) Plus 25% LEO with SRMU
(6) With Centaur upper stage
(7) Plus 10-20% LEO with ASRM
(8) Plus 15% LEO from Cape York
(9) Plus 35% GTO from Cape York
(10) With RCS kick stage
(11) With EUS upper stage

Launch Vehicles, continued

United States

	SCOUT	Enhanced SCOUT	Pegasus	Taurus	Delta II - 6925	Delta II - 7925	Atlas E	Atlas I	Atlas II	Atlas IIA	Atlas IIAS	Titan II SLV	Titan III	Titan IV	Space Shuttle
Year	1979	—	1990	~1992	1989	1990	1974	1990	~1991	~1991	~1993	1988	1989	1989	1981
LEO	600 (270)	1,160 (525)	1,000 (455)	3,200 (1,450)	8,780 (3,990) (3)	11,100 (5,045) (3)	1,800 (820) Polar	12,300 (5,580)	14,100 (6,395)	14,900 (6,760)	18,500 (8,390)	4,200 (1,905) Polar (4)	32,000 (14,515)	39,000 (17,700) (5)	53,700 (24,400) (7)
GTO	120 (54)	240 (110)	275 (125) + PKM	830 (375)	3,190 (1,450)	4,010 (1,820)	—	4,950 (2,250)	5,900 (2,680)	6,200 (2,810)	7,700 (3,490)		11,000 (5,000)+PKM	10,000 (4,540) GEO (6)	13,000 (5,900)
Launch sites	• Wallops • Vandenberg • San Marco	• Wallops • Vandenberg • San Marco	• B-52 aircraft	• Vandenberg	• Cape Canaveral • Vandenberg	• Cape Canaveral • Vandenberg	• Vandenberg	• Cape Canaveral	• Cape Canaveral	• Cape Canaveral	• Cape Canaveral	• Vandenberg	• Cape Canaveral	• Cape Canaveral • Vandenberg	• Kennedy Space Center

Notes:
(1) LEO - low Earth orbit
GTO - geosynchronous transfer orbit
GEO - geosynchronous orbit
PKM - perigee kick motor
(2) Configuration is being finalized
(3) LEO version is designated 6920 or 7920
(4) Plus 95% Polar with 10 solid strap-ons
(5) Plus 25% LEO with SRMU
(6) With Centaur upper stage
(7) Plus 10-20% LEO with ASRM
(8) Plus 15% LEO from Cape York
(9) Plus 35% GTO from Cape York
(10) With RCS kick stage
(11) With EUS upper stage

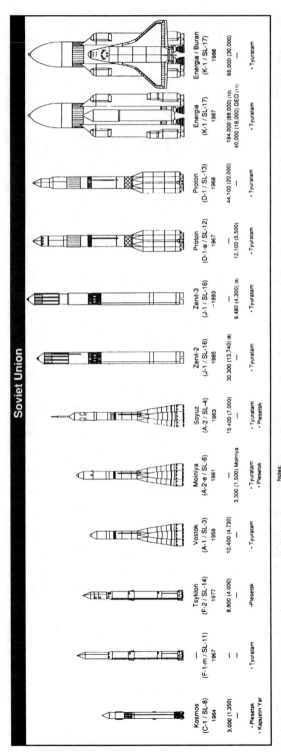

Launch Vehicles, continued

Launch Vehicles, continued

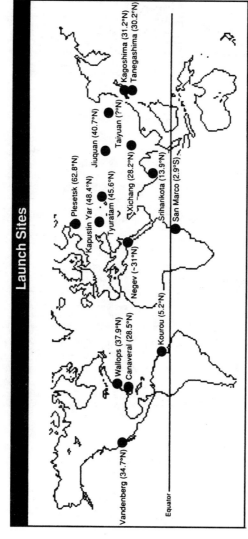

Launch Sites

Ref.: The launch vehicle material appearing on pages 10-5–10-9 is from *International Reference Guide to Space Launch Systems*, 2nd Edition, by S. J. Isakowitz. Copyright © 1995, AIAA, Washington, DC. All rights reserved.

Launch Vehicles, continued

Space Shuttle Launch Configuration

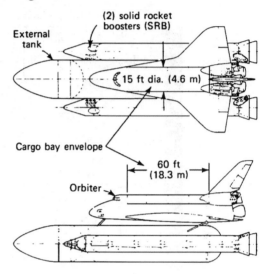

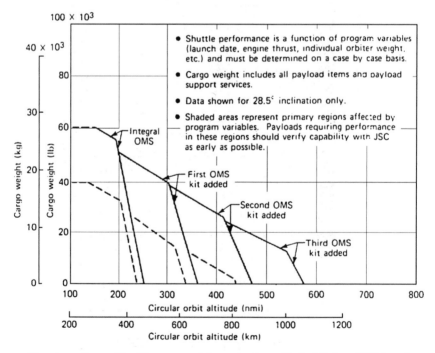

- Shuttle performance is a function of program variables (launch date, engine thrust, individual orbiter weight, etc.) and must be determined on a case by case basis.
- Cargo weight includes all payload items and payload support services.
- Data shown for 28.5° inclination only.
- Shaded areas represent primary regions affected by program variables. Payloads requiring performance in these regions should verify capability with JSC as early as possible.

Near-term Space Shuttle cargo weight vs circular orbital altitude—KSC launch

Space Environment

Space Environment—Hard Vacuum

- Even in low orbit pressure less than best laboratory vacuum
- Many materials, especially polymers, may outgas extensively
  —Change in characteristics
  —Contamination of cold surfaces by recondensing

- Some plating metals—e.g., cadmium—may migrate to cold areas
- Removal of adsorbed O_2 layer can aggravate galling and allow cold welding between surfaces of similar metals—e.g., stainless steel
- Essentially no corrosion
- No convective heat transfer

Atmospheric Characteristics

ALTITUDE VARIATIONS OF DENSITY, TEMPERATURE AND GRAVITY

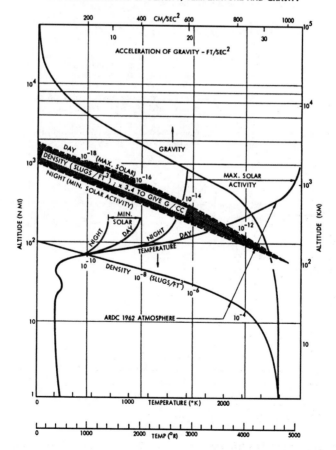

Space Environment, continued

Space Environment—Radiation

- Cosmic Rays
  - —Infrequent, highly energetic
  - —Shielding impractical
  - —Total dose not a problem
  - —Single event upset can be serious

- Solar Flare
  - —Usually not serious for electronics
  - —Human crew requires 2–4 g/cm^2 shielding for average event
  - —Infrequent major event may require 40 g/cm^2 for crew

- Radiation Belts
  - —Charged particles trapped by planetary magnetic field
  - —Earth's Van Allen belts a major problem to both electronics and crew
  - —Jupiter radiation environment very severe

Natural Radiation Environment for Various Orbits

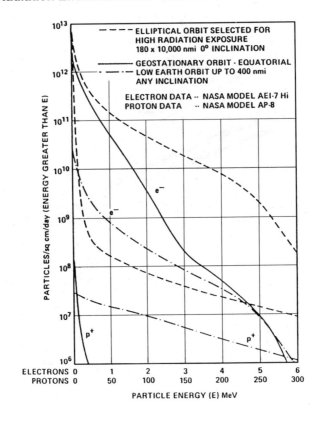

Space Environment, continued

Natural Radiation Environment

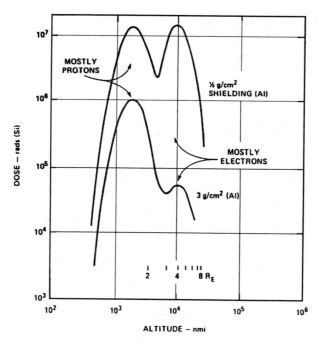

- ELECTRON AND PROTON/
  SPHERICAL SHIELDING
- 10 YEAR MISSION
- CIRCULAR ORBITS/
  0° INCLINATION

Ionized Particles/Charging

- Atomic oxygen attack up to several hundred kilometers
- Oxygen attack highly detrimental to many polymers—e.g., Kapton
- Spacecraft charging
  —Very high voltages generated
  —Affects geosynchronous spacecraft entering eclipse
  —Discharge can cause electronic state changes or permanent damage

Space Environment, continued

The Space Environment Oxygen Atom Flux Variation with Altitude

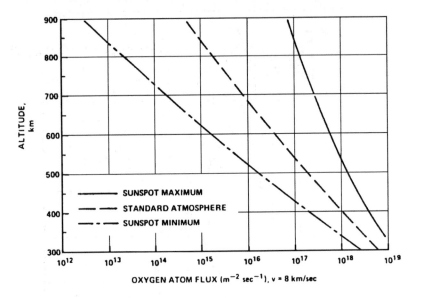

OXYGEN ATOM FLUX (m^{-2} sec^{-1}), v = 8 km/sec

Space Environment—Meteoroid/Debris

Meteoroid

- Large particles very rare
- Most damage pitting/sandblasting
- Near comet environment can be hazardous
- Planetary bodies increase concentration but also shield

Protection

- "Whipple meteor bumper" very effective for high velocity particles
- Large slow particles very difficult

Debris

- Major, growing problem in low-Earth orbit
- Launch debris, explosions, ASAT tests
- Armoring imposes tremendous mass penalty

Space Environment, continued

The Space Environment Micrometeoroid Model

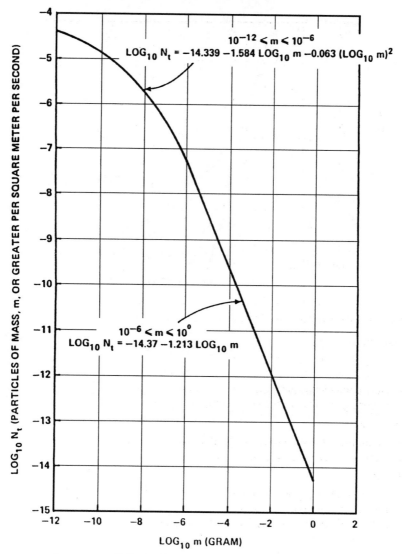

-Average cumulative total meteoroid flux-mass model for 1 A.U.

Space Environment, continued

Space Debris

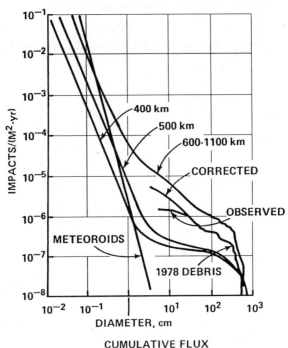

CUMULATIVE FLUX

Shielding in Planet Orbit

Body shielding factor ζ

$$\zeta = \frac{1 + \cos\theta}{2}$$

$$\text{where } \sin\theta = \frac{R}{R + H}$$

ζ = ratio, shielded to unshielded concentration
R = radius of shielding body
H = altitude above surface

Space Environment, continued

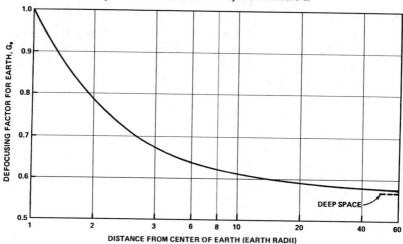

DEFOCUSING FACTOR DUE TO EARTH'S GRAVITY FOR AVERAGE METEOROID VELOCITY OF 20 km/s

The Space Environment Meteoroid and Debris Flux

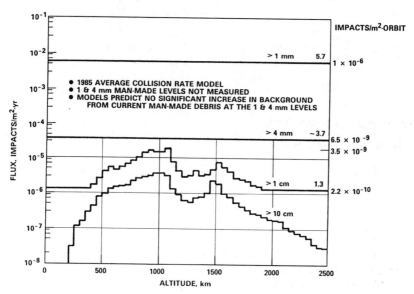

The launch vehicle and space environment material appearing on pages 10-10–10-17 is from *Space Vehicle Design* by M. D. Griffin and J. R. French. Copyright © 1991, AIAA, Washington, DC. All rights reserved.

Space Environment, continued

Properties of the atmospheres of solar system objects [after Pollack and Yung (1980) in which the source references can be found]

Object	$\bar{\rho}^a$	g^a	P_s^a	T_s^a	Major gases[a]	Minor gases[a]	Aerosols[a]
Mercury	5.43	3.95×10^2	$\sim 2 \times 10^{-15}$	440	He (~0.98), H (~0.02)[b]		
Venus	5.25	8.88×10^2	90	730 (~230)	CO_2 (0.96), N_2 (~0.035)	H_2O (20–5000), SO_2 (~150), Ar (20–200), Ne (4–20), CO (50)[c], HCl (0.4)[c], HF (0.01)[c]	Sulphuric acid (~35)
Earth	5.52	9.78×10^2	1	288 (~255)	N_2 (0.77), O_2 (0.21), H_2O (~0.01), Ar (0.0093)	CO_2 (315), Ne (18), He (5.2), Kr (1.1), Xe (0.087), CH_4 (1.5), H_2 (0.5), N_2O (0.3), CO (0.12), NH_3 (0.01), NO_2 (0.001), SO_2 (0.0002), H_2S (0.0002), O_3 (~0.4)	Water (~5), Sulphuric acid (~0.01–0.1)[d], Sulphate, sea salt, Dust, organic (~0.1)[d]
Mars	3.96	3.73×10^2	0.007	218 (~212)	CO_2 (0.95), N_2 (0.027), Ar (0.016)	O_2 (1300), CO (700), H_2O (~300), Ne (2.5), Kr (0.3), Xe (0.08), O_3 (~0.1)	Water ice (~1)[c], Dust (~0.1–10)[c], CO_2 ice (?)[c]
Moon	3.34	1.62×10^2	$\sim 2 \times 10^{-14}$	274	Ne (~0.4), Ar (~0.4), He (~0.2)	—	—

(continued)

[a] Reading from left to right, the variables are the object's mean density (g cm^{-3}); acceleration of gravity (cm s^{-2}); surface pressure (bar); surface temperature (K), the numbers in parentheses are values of effective temperature; major gas species; the numbers in parentheses are volume mixing ratios; minor gas species, the numbers in parentheses are fractional abundance by number in units of ppm; aerosol species, the numbers in parentheses are typical values of the aerosols' optical depth in the visible. The numbers in this table were derived from the sources cited in Pollack and Yung (1980). [b] These mixing ratios refer to typical values at the surface. [c] These mixing ratios pertain to the region above the cloud tops. [d] The sulphuric acid aerosol resides in the lower stratosphere, while the sulphate, etc., aerosols are found in the troposphere, especially in the bottom boundary layer. [e] The ice clouds are found preferentially above the winter polar regions. Dust particles are present over the entire globe. [f] These mixing ratios pertain to the stratosphere. Fink et al. (1980) have reported the detection of a CH_4 atmosphere on Pluto. Ref.: *The Origin and Evolution of Planetary Atmospheres* by A. Henderson-Sellers. Copyright © 1983, IOP Publishing Limited, Bristol, England. Reprinted by permission of IOP Publishing Limited.

Space Environment, continued

Properties of the atmospheres of solar system objects, continued

Object	$\bar{\rho}^a$	g^a	$P_s{}^a$	$T_s{}^a$	Major gases[a]	Minor gases[a]	Aerosols[a]
Jupiter	1.34	2.32×10^3	$\gg 100^f$	(129)	H_2 (~0.89), He (~0.11)	HD (20), CH_4 (~2000), NH_3 (~200), H_2O (1?), C_2H_6 (~5^f, CO (0.002), GeH_4 (0.0007), HCN (0.1), C_2H_2 (~0.02)^f, PH_3 (0.4)	Stratospheric "smog" (~0.1), Ammonia ice (~1), Ammonium hydrosulphide (~1), Water (~10)
Saturn	0.68	8.77×10^2	$\gg 100^f$	(97)	H_2 (~0.89), He (~0.11)	CH_4 (3000), NH_3 (~200), C_2H_6 (~2)^f	Same aerosol layers as for Jupiter
Uranus	1.55	9.46×10^2	$\gg 100^f$	(58)	H_2 (~0.89),	CH_4, He (~0.11)	Same aerosol layers as for Jupiter, but thinner smog layer, plus possibly methane ice
Neptune	2.23	1.37×10^3	$\gg 100^f$	(56)	H_2 (~0.89),	CH_4, He (~0.11)	Same aerosol layers as for Jupiter, plus possibly methane ice
Titan	~1.4	$\sim 1.25 \times 10^2$	$2 \times 10^{-2} \rightarrow \sim 1$	~85	CH_4 (0.1–1)	C_2H_6 (~2)	Stratospheric "smog" (~10)
Io	3.52	1.79×10^2	$\sim 1 \times 10^{-10}$	~110	SO_2 (~1)	—	—

[a] Reading from left to right, the variables are the object's mean density (g cm^{-3}); acceleration of gravity (cm s^{-2}); surface pressure (bar); surface temperature (K), the numbers in parentheses are values of effective temperature; major gas species, the numbers in parentheses are volume mixing ratios; minor gas species, the numbers in parentheses are fractional abundance by number in units of ppm; aerosol species, the numbers in parentheses are typical values of the aerosols' optical depth in the visible. The numbers in this table were derived from the sources cited in Pollack and Yung (1980). [b] These mixing ratios refer to typical values at the surface. [c] These mixing ratios pertain to the region above the cloud tops. [d] The sulphuric acid aerosol resides in the lower stratosphere, while the sulphate, etc., aerosols are found in the troposphere, especially in the bottom boundary layer. [e] The ice clouds are found preferentially above the winter polar regions. Dust particles are present over the entire globe. [f] These mixing ratios pertain to the stratosphere. Fink et al. (1980) have reported the detection of a CH_4 atmosphere on Pluto. Ref.: *The Origin and Evolution of planetary Atmospheres* by A. Henderson-Sellers. Copyright © 1983, IOP Publishing Limited, Bristol, England. Reprinted by permission of IOP Publishing Limited.

Space Environment, continued

Atmospheric Characteristics—The Solar System

GENERAL DIRECTION OF REVOLUTION
ABOUT SUN - COUNTERCLOCKWISE
VIEWED FROM POINT NORTH OF ECLIPTIC

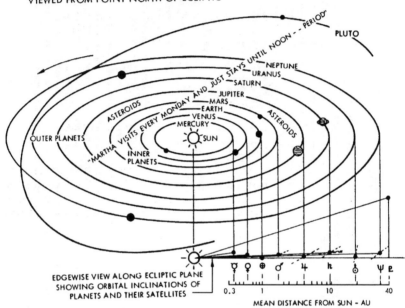

EDGEWISE VIEW ALONG ECLIPTIC PLANE
SHOWING ORBITAL INCLINATIONS OF
PLANETS AND THEIR SATELLITES

MEAN DISTANCE FROM SUN - AU

Earth Data

Law of Gravitation: $F = Gm_1m_2/r^2 =$ force of attraction between two bodies, masses m_1 and m_2, separated by a distance r.

$$G = 6.664 \times 10^{-8} \text{ dyne-cm}^2/\text{g}^2$$

Mean equatorial radius of Earth = 6378.1 km = 3963.1 st mile
                                                      = 3443.9 n mile
Polar radius of Earth = 6356.9 km = 3950.0 st mile
                                              = 3430.0 n mile
Equatorial radius of Earth = 6378.4 km
                                           = 3963.3 st mile
                                           = 3441.6 n mile

Space Environment, continued

Mass of Earth $= 5.983 \times 10^{24}$ kg $= 13.22 \times 10^{24}$ lb

Average density of Earth $= 5.522$ g/cm^3
$$= 344.7 \text{ lb/ft}^3$$

Mean linear velocity of Earth in orbit $= 29.79$ km/s
$$= 18.50 \text{ st mile/s}$$

Mean linear surface rotation velocity of Earth at equator $= 0.465$ km/s
$$= 0.289 \text{ st mile/s}$$

1 deg of latitude $= 60$ n mile $= 69$ st mile

$1'$ of arc on Earth's surface at equator $= 1$ n mile

Orbital Mechanics

Delta V Requirements for Altitude and Inclination Changes

Purpose of the Chart

The following chart may be used to determine the change in velocity required to change the altitude or inclination of a low-Earth orbit. The Delta V required can then be translated into the mass of propellant required to complete the maneuver. Note that the graphs for both cases are linear for the range between 200 and 1000 km.

Equations

$$|\Delta V|a = [398{,}600.8/R_f]^{0.5} - [398{,}600.8/R_i]^{0.5} * 1000 \text{ m/s}$$

$$|\Delta V|i = 2 * V_i * \sin(\Delta i/2) * 1000 \text{ m/s}$$

where

$|\Delta V|a = \Delta V$ required for a change in altitude, m/s
$|\Delta V|i = \Delta V$ required for a change in inclination, m/s
R_i $\quad$ = initial radius from the center of Earth to s/c, m/s
R_f $\quad$ = final or desired radius, m/s
V_i $\quad$ = orbital velocity at beginning of maneuver, km/s
Δ_i $\quad$ = desired change in inclination, deg
R_e $\quad$ = radius of Earth, 6378.135, km

Assumptions

Standard two-body conditions
Low-Earth, circular orbit

Orbital Mechanics, continued

Delta V for Altitude and Inclination Changes

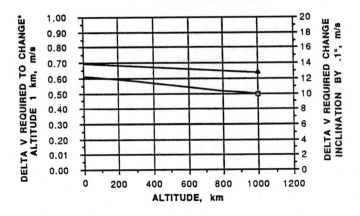

* For Hohmann Transfer

LEGEND
- -▣- Delta V for 1 km altitude change
- -▲- Delta V for 0.1° inclination change

Simplified ΔV Calculation for Orbit Transfer

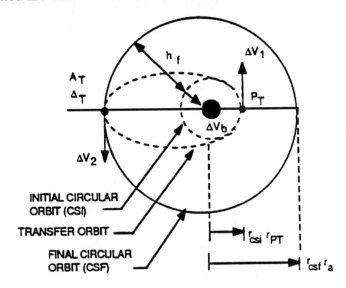

Orbital Mechanics, continued

The required velocity for orbit injection can be approximated by

$$\Delta V_{Total} = \Delta V_b + \Delta V_{sc}$$

where

ΔV_b = change in velocity provided by the booster
$\Delta V_{sc} = \Delta V_1 + \Delta V_2 + \Delta V_s$ = change in velocity provided by the spacecraft and/or the upper stage
ΔV_1, ΔV_2 are shown in the diagram
ΔV_s = change in velocity to alter the inclination of the orbit

Hohmann Transfer (ΔV_1, ΔV_2)—changes size and shape of the orbit, but not the inclination
Simple plane change (ΔV_s)—changes the inclination by Δi

Calculations

$$r_{CSI} = r_{PT} = 6378.135 + h_i \text{ km}$$

$$r_{CSF} = r_{AT} = 6378.135 + h_f \text{ km}$$

$$2a_T = r_{CSI} + r_{CSF} \text{ km}$$

$$\varepsilon_T = \frac{-398,600.8}{2a_T} \text{ km/s}^2$$

$$V_{CSI} = \sqrt{\frac{398,600.8}{r}} \text{ km/s}$$

$$V_{PT} = \sqrt{2\left(\varepsilon_T + \frac{398,600.8}{r_{CSI}}\right)} \text{ km/s}$$

$$V_{AT} = \sqrt{2\left(\varepsilon_{T_{CSI}} + \frac{398,600.8}{r_{CSF}}\right)} \text{ km/s}$$

$$V = \sqrt{\frac{398,600.8}{r_{CSF}}} \text{ km/s}$$

$\Delta V_1 = |V_{PT} - V_{CSI}|$ km/s   (typically provided by apogee kick motor)
$\Delta V_2 = |V_{CSF} - V_{AT}|_{\Delta i}$ km/s   (typically provided by apogee kick motor)

> Check
> $$\Delta V_1 + \Delta V_2 = V_{CSI} - V_{CSF}$$

$$\Delta V_S = 2V_{CSF} \sin \frac{\Delta i}{2} \qquad \Delta i = \text{change in inclination}$$

$$\Delta V_{S/C} = \Delta V_1 + \Delta V_2 + \Delta V_S$$

Orbital Mechanics, continued

Elliptical Orbit Parameters

$$\tan \frac{\theta}{2} = \sqrt{\frac{1+e}{1-e}} \tan \frac{E}{2}$$

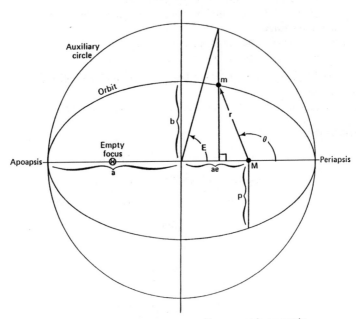

E = eccentric anomaly

Circle	Ellipse	Parabola	Hyperbola
$a = b > 0$	$a > b > 0$	$a = \infty$	$a < 0$
$e = 0$	$e^2 = 1 - \frac{b^2}{a^2}$	$e = 1$	$e^2 = 1 + \frac{b^2}{a^2}$

$$r = \frac{p}{1+e\cos\theta} = \frac{r_p(1+e)}{1+e\cos\theta} = \frac{a(1-e^2)}{1+e\cos\theta}$$

$$p = a(1 - e^2) = h^2/\mu$$

$$r_p = a(1 - e) = \text{periapsis radius}$$

$$r_a = a(1 + e) = \text{apoapsis radius}$$

$$a = \frac{r_a + r_p}{2} \qquad V_{\text{CIRC}}^2 = \frac{\mu}{r}$$

$$e = \frac{r_a - r_p}{r_a + r_p} \qquad V_{\text{ESC}}^2 = \frac{2\mu}{r}$$

Orbital Mechanics, continued

$$V^2 = \mu\left(\frac{2}{r} - \frac{1}{a}\right)$$

$$E_t = \frac{V^2}{2} - \frac{\mu}{r} = -\frac{\mu}{2a}$$

$$e^2 = 1 + 2E_t\frac{h^2}{\mu^2}$$

$$E = -e\sin E - n(t - t_p) = 0 \quad \text{(Kepler's equation)}$$

where

t_p = time of periapsis
n = mean motion = $\sqrt{\mu/a^3}$
$M \overset{\Delta}{=} n(t - t_p)$ = mean anomaly
E = eccentric motion
     for $E = 2\pi$, $\gamma = t - t_p = 2\pi\sqrt{a^3/\mu}$
     for small e, $\theta \cong M + 2e\sin M + (5e^2/4)\sin 2M$

Orbital Elements

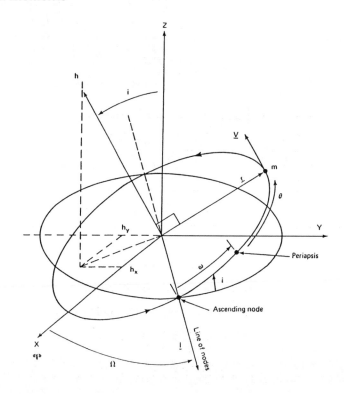

Orbital Mechanics, continued

Hyperbolic Orbit Geometry

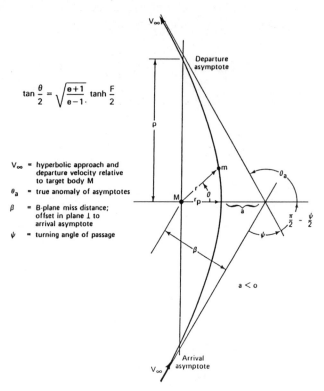

$$\tan \frac{\theta}{2} = \sqrt{\frac{e+1}{e-1}} \ \tanh \frac{F}{2}$$

V_∞ = hyperbolic approach and departure velocity relative to target body M

θ_a = true anomaly of asymptotes

β = B-plane miss distance; offset in plane $\perp$ to arrival asymptote

ψ = turning angle of passage

$$(e \sinh F) - F - n(t - t_p) = 0$$

$$n = \sqrt{\mu/(-a)^3}$$

Key point: Hyperbolic orbit must be oriented in external frame to account for arrival/departure conditions.

At "infinity," the asymptote conditions are as follows:

$$\Theta_a = \cos^{-1}(-1/e)$$

$$V_\infty^2 = 2E_t = (-\mu/a)$$

$$h = \beta V_\infty$$

Key point: Hyperbolic passage alters $\underline{V}_\infty$, but not V_∞.

$$\Psi = 2\sin^{-1}(1/e) = 2\Theta_a - \pi = \text{turning angle}$$

$$\Delta V = 2V_\infty \sin \Psi/2 = 2V_\infty/e$$

The orbital mechanics material appearing on pages 10-24–10-26 is from *Space Vehicle Design* by M. D. Griffin and J. R. French. Copyright © 1991, AIAA, Washington, DC. All rights reserved.

Orbital Mechanics, continued

This is the basis for "gravity assist" maneuvers.

$$e = 1 + \left(r_p V_\infty^2/\mu\right) = \sqrt{\left[1 + \left(\beta V_\infty^2/\mu\right)^2\right]}$$

$$\beta = r_p\sqrt{\left[1 + \left(2\mu/r_p V_\infty^2\right)\right]}$$

$$\left(r_p/\beta\right) = \left(-\mu/\beta V_\infty^2\right) = \sqrt{\left\{1 + \left(\mu/\beta V_\infty^2\right)^2\right\}}$$

Satellite Lifetime vs Altitude

Purpose of the Chart

The following chart may be used to determine the approximate lifetime of a satellite based on aerodynamic drag acting on the spacecraft. End of life was arbitrarily defined as a loss of 50 km of altitude, and no additional propulsion was assumed.

Equations

$$F = m * a$$

$$F_d = M_{SC} * \Delta V/\Delta t = \tfrac{1}{2} * \rho * C_d * V^2 * A_f$$

Solving for Δt in Years

$$\text{Satellite, years} = [M_{SC}/(C_d * A_f)] * \Delta V/(\rho * V^2 * 3600 * 24 * 365)$$

where

F_d = force due to atmospheric drag, N
M_{SC} = mass of the spacecraft, kg
ΔV = change in velocity, km/s
Δt = increment in time, years in this case
ρ = atmospheric density, kg/m^3
C_d = coefficient of drag, assumed 2.0
V = orbital velocity, km/s
A_f = projected area of s/c perpendicular to velocity, m^2
C_B = ballistic coefficient = $M_{SC}/(C_d * A_f)$, kg/m^2

Assumptions

Circular orbit
Standard two-body conditions

The orbital mechanics material appearing on pages 10-27–10-32 is from *Spacecraft Systems Design Handbook* by W. J. Larson. Copyright © 1988, Kluwer Publishing. Reproduced with permission of Kluwer Publishing.

Orbital Mechanics, continued

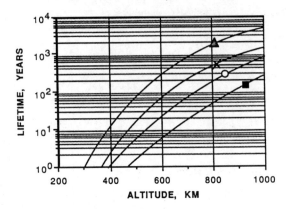

END OF LIFE FOR THE SATELLITE IS
DEFINED AS LOSS OF 50 KM IN ORBITAL
ALTITUDE.

LEGEND

■ BC = 50 kg/m^2, T = 1200 K
O BC = 200 kg/m^2, T = 1200 K
X BC = 50 kg/m^2, T = 800 K
▲ BC = 200 kg/m^2, T = 800 K

Nodal Regression and Perigee Rotation

Purpose of the Chart

The oblateness of the Earth can cause fairly large secular changes in the location of the ascending node and the location of perigee. This chart yields the approximate nodal regression (movement of the longitude of ascending node) and the rotation of perigee for the specified circular orbit.

Equations

$$\dot{\Omega} = (-1.11/365) * \cos(i) / [(a/R_e)^{3.5} * (1 - e^2)^2]$$

$$\dot{\omega} = (1.11/365) * (2 - 2.5 * \sin^2(i)) / [(a/R_e)^{3.5} * (1 - e^2)^2]$$

where

$\dot{\Omega}$ = nodal regression, (inertial) deg/mean solar day
$\dot{\omega}$ = rotation of perigee, (inertial) deg/mean solar day
i = orbit inclination, deg
a = semimajor axis, km, $a = R$ for circular orbit
R_e = radius of Earth, 6378.135 km
e = eccentricity of orbit

Orbital Mechanics, continued

Assumptions

Standard two-body conditions
Low-Earth orbit

Nodal Regression for Circular Orbits

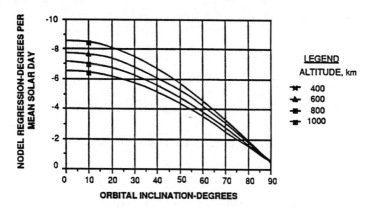

NODES MOVE:
WESTWARD IF ORBIT INCLINATION IS BETWEEN
0° AND 90° (DIRECT ORBIT)
EASTWARD IF ORBIT INCLINATION IS BETWEEN
90° AND 180° (RETROGRADE ORBIT)

Rotation of Perigee

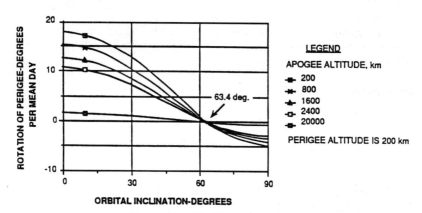

Orbital Mechanics, continued

Inclination for Sun-Synchronous Orbits

Purpose of the Chart

The oblateness of the Earth can cause fairly large secular changes in the location of the longitude of the ascending node. If the equations below are solved for inclination and the nodal regression rate is set to the rotation rate of the Earth about the sun, about 1 per day, the following graphs will result. These charts may be used to approximate the inclination of sun-synchronous, circular, or elliptical orbits.

Equations

$$\dot{\Omega} = (-1.11/365) * \cos(i) / \left[(a/R_e)^{3.5} * (1 - e^2)^2 \right]$$

$$\dot{\omega} = (1.11/365) * (2 - 2.5 * \sin^2(i)) / \left[(a/R_e)^{3.5} * (1 - e^2)^2 \right]$$

where

$\dot{\Omega}$ = nodal regression, deg/mean solar day
$\dot{\omega}$ = rotation of perigee, deg/mean solar day
i = orbit inclination, deg
a = semimajor axis, km, $a = R$ for circular orbit
R_e = radius of Earth, 6378.135 km
e = eccentricity of orbit

Assumptions

Standard two-body conditions
Low-Earth orbit

Orbital Mechanics, continued

Circular

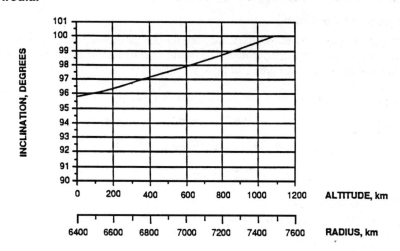

Elliptical

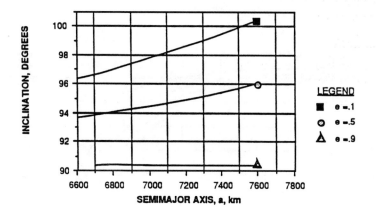

Orbital Mechanics, continued

Eclipse Time vs Sun Angle

Purpose of the Chart

The following chart may be used to determine the duration of satellite eclipse, if any, for various circular orbit altitudes. Given the sun/orbit angle β and the altitude, the eclipse duration in minutes can be approximated.

Equations

$$T_e = P/\pi * \cos^{-1}\left[\left(1 - R_e^2/R^2\right)^{0.5}/\cos(\beta)\right]$$

$$P = 2*\pi*[R^3/398,600.8]^{0.5}*60$$

where

T_e = eclipse time, min
P = orbital period, min
R_e = radius of Earth, 6378.135 km
R = distance from the center of Earth to the s/c, km
β = sun/orbit angle, deg

Assumptions

Standard two-body conditions
Low-Earth orbit

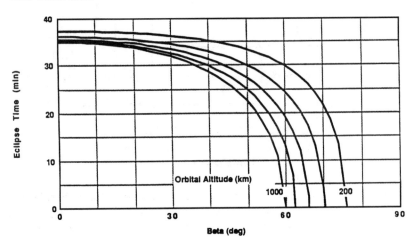

Propulsion Systems

Types of Propulsion Systems

Cold Gas

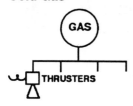

Characteristics:

$I_{sp} \cong 50\,s$

$I_{sp} \cong K\sqrt{\dfrac{T}{M}}$

$F < 1\,lb$

Monopropellants

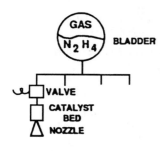

Characteristics:

$I_{sp} \cong 225\,s$

$1 < F < 600\,lb$

Bipropellants

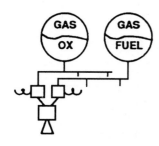

Characteristics:

$I_{sp} \cong 310–460\,s$

Propulsion Systems, continued

Solids

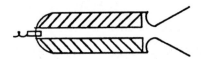

Characteristics:

$I_{sp} \cong 250\text{--}290$ s

Dual Mode Propulsion System

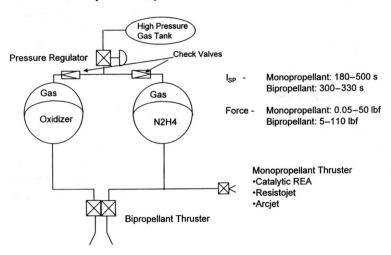

I_{SP} -   Monopropellant: 180–500 s
      Bipropellant: 300–330 s

Force -   Monopropellant: 0.05–50 lbf
       Bipropellant: 5–110 lbf

Monopropellant Thruster
•Catalytic REA
•Resistojet
•Arcjet

Resistojet (Electrothermal Hydrazine Thruster) System

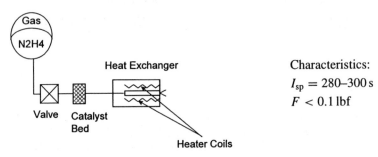

Characteristics:

$I_{sp} = 280\text{--}300$ s

$F < 0.1$ lbf

Propulsion Systems, continued

Summary of Key Equations for Ideal Rockets

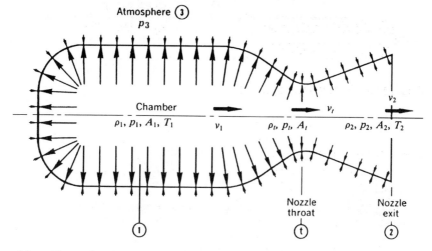

Note: Four subscripts are used above (1, 2, 3, t). They are shown in circles and refer to the specific locations. Thus, p_1 is the chamber pressure, p_2 is the nozzle exit pressure, p_3 is the external fluid of atmospheric pressure, and p_t is the nozzle throat pressure. T_1 is the combustion chamber absolute temperature.

Parameter	Equations
Average exhaust velocity, v_2 (m/s or ft/s) (assume that $v_1 = 0$)	$v_2 = c - (p_2 - p_3)A_2/\dot{m}$ When $p_2 = p_3$, $v_2 = c$ $v_2 = \sqrt{[2k/(k-1)]RT_1\left[1 - (p_2/p_1)^{(k-1)/k}\right]}$ $\quad = \sqrt{2(h_1 - h_2)}$
Effective exhaust velocity, c (m/s or ft/s) Thrust, F (N or lbf)	$c = c^*C_F = F/\dot{m} = I_{sp}g_0$ $c = v_2 + (p_2 - p_3)A_2/\dot{m}$ $F = c\dot{m} = cm_p/t_b$ $F = C_F p_1 A_t$ $F = \dot{m}v_2 + (p_2 - p_3)A_2$ $F = \dot{m}I_{sp}/g_0$
Characteristic exhaust velocity, c^* (m/s or ft/s)	$c^* = c/C_F = p_1 A_t/\dot{m}$ $c^* = \dfrac{\sqrt{kRT_1}}{k\sqrt{[2/(k+1)]^{(k+1)(k-1)}}}$ $c^* = I_{sp}g_0/C_F$

(continued)

Propulsion Systems, continued

Parameter	Equations
Thrust coefficient, C_F (dimensionless)	$C_F = c/c^* = F/(p_1 A_t)$

$$C_F = \sqrt{\frac{2k^2}{k-1}\left(\frac{2}{k+1}\right)^{(k+1)/(k-1)}\left[1 - \left(\frac{p_2}{p_1}\right)^{(k-1)/k}\right]}$$
$$+ \frac{p_2 - p_3}{p_1}\frac{A_2}{A_t}$$

Parameter	Equations
Mass flow rate, $\dot{m}$ (kg/s or lb/s)	$\dot{m} = \dfrac{A_t v_t}{V_t} = p_1 A_1/c^*$

$$= A_t p_1 \frac{k\sqrt{[2/(k+1)]^{(k+1)/(k-1)}}}{\sqrt{kRT_1}}$$

Total impulse, I_t [N-s or lbf-s]	$I_t = \displaystyle\int_0^{t_b} F\,dt$ $= F t_b$ (F constant over t_b)
Specific impulse, I_{sp} [N-s/(kg × 9.8066 m/s²) or lbf-s/lbm or s]	$I_{sp} = c/g_0 = c^* C_F/g_0$ $I_{sp} = F/\dot{m} g_0 = F/\dot{w}$ $I_{sp} = v_2/g_0 + (p_2 - p_3)A_2/(\dot{m} g_0)$ $I_{sp} = I_t/(m_p g_0) = I_t/w_p$
Propellant mass fraction (dimensionless)	$\zeta = m_p/m_0 = \dfrac{m_0 - m_f}{m_0}$ $\zeta = 1 - \text{MR}$
Mass ratio of vehicle or stage, MR (dimensionless)	$\text{MR} = \dfrac{m_f}{m_0} = \dfrac{m_0 - m_p}{m_0}$ $= m_f/(m_f + m_p)$ $m_0 = m_f + m_p$
Vehicle velocity increase in gravity free vacuum, Δv (m/s or ft/s) (assume that $v_0 = 0$)	$\Delta v = -c \ln \text{MR} = +c \ln \dfrac{m_0}{m_f}$ $= c \ln m_0/(m_0 - m_p)$ $= c \ln(m_p + m_f)/m_f$ $= I_{sp} g_0 \ln \text{MR}$
Propellant mass flow rate, $\dot{m}$ (kg/s or lb/s)	$\dot{m} = Av/V = A_1 v_1/V_1$ $= A_t v_t/V_t = A_2 V_2/V_2$ $\dot{m} = F/c = p_1 A_t/c^*$

$$\dot{m} = p_1 A_t k \sqrt{\frac{[2/(k+1)]^{(k+1)/(k-1)}}{\sqrt{kRT_1}}}$$
$$\dot{m} = m_p/t_b$$

(continued)

Propulsion Systems, continued

Parameter	Equations
Mach number, M (dimensionless)	$M = v/a$ $= v/\sqrt{kRT}$ At throat $v = a$ and $M = 1.0$
Nozzle area ratio, ϵ	$\epsilon = A_2/A_t$ $\epsilon = \dfrac{1}{M_2}\left[\dfrac{1 + \frac{k-1}{2}M_2^2}{1 + \frac{k-1}{2}}\right]^{(k+1)/(k-1)}$
Isentropic flow relationships for stagnation and free-stream conditions	$T_0/T = (p_0/p)^{(k-1)/k} = (V/V_0)^{k-1}$ $T_x/T_y = (p_x/p_y)^{(k-1)/k} = (V_x/V_y)^{k-1}$

where

R	$= R'/\mathfrak{M}$
R'	= universal gas constant
	= 8314 J/kg mol-K (1544 ft-lb/mol-°R)
$\mathfrak{M}$	= molecular weight of reaction gases, kg (lbm)
k	= ratio of specific heats $= c_p/c_v$
c_p	= specific heat at constant pressure
c_v	= specific heat at constant volume
T	= temperature, K (°R)
ρ	= density, kg/m³ (lbm/ft³)
V	= specific volume, m³/kg (ft³/lbm)
p	= pressure, N/m² (lb/ft²)
h	= enthalpy $= c_p T$
F	= thrust, N (lbf)
g_0	= acceleration due to gravity on Earth = 9.81 m/s (32.2 ft/s²)
t_b	= total burn time
m_p	= total propellant mass
w_p	= total propellant weight
A	= cross-sectional area
m_0	= mass at start of burn
m_f	= mass at end of burn
v	= flow velocity
a	= speed of sound
T_0, k_0, V_0	= stagnation conditions

Propulsion Systems, continued

Summary of Key Equations for Solid Rocket Motors

Mass flow rate, kg/s (lbm/s):

$$\dot{m} = A_b r \rho_b$$

$$= \frac{d}{dt}(\rho_1 V_1) + A_t p_1 \sqrt{\frac{k}{RT_1}\left(\frac{2}{k+1}\right)^{(k+1)/(k-1)}}$$

Burn rate r, in./s:

$$r = a p_1^n$$

Propellant mass m_p, kg (lbm):

$$m_p = \int \dot{m}\,dt = \rho_b \int A_b r\,dt$$

Assuming:

$$\frac{d}{dt}(\rho_t V_c) \ll A_t p_1 \sqrt{\frac{k}{RT_1}\left(\frac{2}{k+1}\right)^{(k+1)/(k-1)}}$$

(Rate of change of gas mass in motor cavity is small relative to mass flow through the nozzle.)
     Then,

$$\frac{A_b}{A_t} = K \begin{bmatrix} \text{Ratio of burning area} \\ \text{to nozzle throat area} \end{bmatrix}$$

$$= \frac{p_1^{(1-n)}\sqrt{k[2/(k+1)]^{(k+1)/(k-1)}}}{\rho_b a \sqrt{RT_1}}$$

It follows:

$$p_1 = [K \rho_b a c^*]^{1/(1-n)}$$

where

A_b = burn area of propellant grain
ρ_b = solid propellant density prior to motor start
V_1 = combustion chamber gas volume
a = burn rate coefficient
n = burn rate exponent

Propulsion Systems, continued

Typical Performance of Propulsion Types

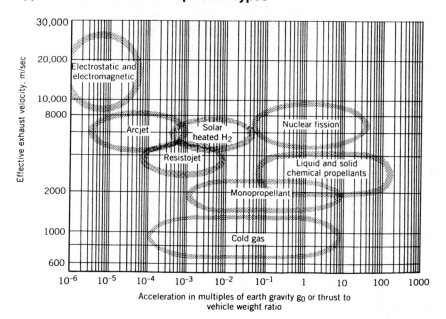

Propulsion Systems, continued

Ranges of Typical Performance Parameters for Various Rocket Propulsion Systems

Engine type	Specific impulse,[a] s	Maximum temperature, °C	Thrust-to-weight ratio[b]	Propulsion duration	Specific power,[c] kW/kg	Typical working fluid	Status of technology
Chemical-solid or liquid bipropellant	200–410	2500–4100	10^{-2}–100	Seconds to a few minutes	10^{-1}–10^3	Liquid or solid propellants	Flight proven
Liquid monopropellant	180–223	600–800	10^{-1}–10^{-2}	Seconds to minutes	0.02–200	N_2H_4	Flight proven
Nuclear fission	500–860	2700	10^{-2}–30	Same	10^{-1}–10^3	H_2	Development was stopped
Resistojet	150–300	2700	10^{-2}–10^{-4}	Days	10^{-3}–10^{-1}	H_2, N_2H_4	Flight tested
Arc heating—electrothermal	280–1200	5500	10^{-4}–10^{-2}	Days	10^{-3}–1	N_2H_4, H_2, NH_3	Flight tested
Electromagnetic	1200–6000	—	10^{-6}–10^{-4}	Weeks	10^{-3}–1	H_2	Several have flown
Ion—electrostatic	1200–5000	—	10^{-6}–10^{-4}	Months	10^{-3}–1	Xe	Several have flown
Solar heating	400–700	1300	10^{-3}–10^{-2}	Days	10^{-2}–1	H_2	Not yet flight tested

[a] At p_1 = 1000 psia and optimum gas expansion at sea level ($p_2 = p_3$ = 14.7 psia).
[b] Ratio of thrust force to full propulsion system sea level weight (with propellants, but without payload).
[c] Kinetic energy per unit exhaust mass flow.

Propulsion Systems, continued
Properties of Gaseous Propellants Used for Auxiliary Propulsion

Propellant	Molecular mass	Density,[a] lb/ft^3	Theoretical specific impulse,[b] s
Hydrogen	2.0	1.77	284
Helium	4.0	3.54	179
Methane	16.0	14.1	114
Nitrogen	28.0	24.7	76
Air	28.9	25.5	74
Argon	39.9	35.3	57
Krypton	83.8	74.1	50

[a] At 5000 psia and 20°C.
[b] In vacuum with nozzle area ratio of 50:1 and initial temperature of 20°C.

Engine type	η_{int}	I_{sp}	v_2, m/s	$\dot{m}$, kg/s	Power input, kW
Chemical rocket	0.50	300	2,352	0.0425	117
Nuclear fission	0.50	800	6,860	0.0145	682
Arc–electrothermal	0.50	1200	10,780	0.0093	1351
Ion electrostatic	0.90	5000	49,000	0.0020	2668

$$\eta_{int} = \frac{\text{power of the jet (output)}}{\text{power input}}$$

$$= \frac{\frac{1}{2}\dot{m}C^2}{P} = \frac{F I_{sp} g_0}{2P}$$

Propulsion Systems, continued

Theoretical Performance of Liquid Rocket Propellant Combinations[a]

Oxidizer	Fuel	Mixture ratio		Average specific gravity, g/cm³	Chamber temp, K	c^*, m/s	$\mathfrak{M}$, kg/mol	I_{sp}, s	k
		By mass	By volume						
Oxygen	75% Ethyl alcohol	1.30	0.98	1.00	3177	1641	23.4	267	1.22
		1.43	1.08	1.01	3230	1670	24.1	279	
	Hydrazine	0.74	0.66	1.06	3285	1871	18.3	301	1.25
		0.90	0.80	1.07	3404	1892	19.3	313	
	Hydrogen	3.40	0.21	0.26	2959	2428	8.9	387	1.26
		4.02	0.25	0.28	2999	2432	10.0	390	
	RP-1	2.24	1.59	1.01	3571	1774	21.9	286	1.24
		2.56	1.82	1.02	3677	1800	23.3	300	
	UDMH	1.39	0.96	0.96	3542	1835	19.8	295	1.25
		1.65	1.14	0.98	3594	1864	21.3	310	

(continued)

[a]Combustion chamber pressure—1000 psia (6895 kN/m²). Nozzle exit pressure—14.7 psia (1 atm). Optimum nozzle expansion ratio. Adiabatic combustion and isentropic expansion of ideal gas. Compositions expressed in mass percent. The density at the boiling point was used for those oxidizers or fuels that boil below 20°C at 1 atm pressure. For every propellant combination, there are two sets of values listed; the upper line refers to frozen equilibrium, the lower line to shifting equilibrium.

Propulsion Systems, continued

Theoretical Performance of Liquid Rocket Propellant Combinations[a], continued

Oxidizer	Fuel	Mixture ratio		Average specific gravity, g/cm³	Chamber temp, K	c^*, m/s	$\mathfrak{M}$, kg/mol	I_{sp}, s	k
		By mass	By volume						
Fluorine	Hydrazine	1.83	1.22	1.29	4553	2128	18.5	334	1.33
		2.30	1.54	1.31	4713	2208	19.4	363	
	Hydrogen	4.54	0.21	0.33	3080	2534	8.9	398	1.33
		7.60	0.35	0.45	3900	2549	11.8	410	
Nitrogen tetroxide	Hydrazine	1.08	0.75	1.20	3258	1765	19.5	283	1.26
		1.34	0.93	1.22	3152	1782	20.9	292	
	50% UDMH⁻ 50% hydrazine	1.62	1.01	1.18	3242	1652	21.0	278	1.24
		2.00	1.24	1.21	3372	1711	22.6	288	
Red fuming nitric acid	RP-1	4.1	2.12	1.35	3175	1594	24.6	258	1.22
		4.8	2.48	1.33	3230	1609	25.8	268	
	50% UDMH- 50% hydrazine	1.73	1.00	1.23	2997	1682	20.6	272	1.22
		2.20	1.26	1.27	3172	1701	22.4	279	

[a]Combusion chamber pressure—1000 psia (6895 kN/m²). Nozzle exit pressure—14.7 psia (1 atm). Optimum nozzle expansion ratio. Adiabatic combustion and isentropic expansion of ideal gas. Compositions expressed in mass percent. The density at the boiling point was used for those oxidizers or fuels that boil below 20°C at 1 atm pressure. For every propellant combination, there are two sets of values listed; the upper line refers to frozen equilibrium, the lower line to shifting equilibrium.

Propulsion Systems, continued

Typical Turbopump Feed System Cycles for Liquid-Propellant Rocket Engines

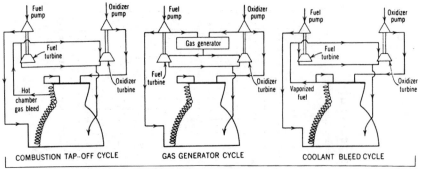

COMBUSTION TAP-OFF CYCLE   GAS GENERATOR CYCLE   COOLANT BLEED CYCLE

Open Cycles

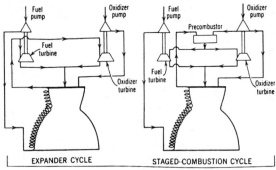

EXPANDER CYCLE   STAGED-COMBUSTION CYCLE

Closed Cycles

Propulsion Systems, continued

Burning Rate vs Chamber Pressure for Some Typical Propellants at Several Propellant Temperatures

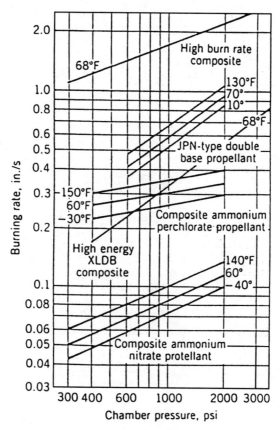

Burn rate temperature sensitivity at constant pressure:

$$\sigma_p = 1/r \left(\frac{\delta r}{\delta T} \right)_p$$

Burn rate temperature sensitivity at constant K:

$$\pi_K = 1/p_c \left(\frac{\delta r}{\delta T} \right)_K$$

Generally,

$$0.08\% < \sigma_p < 0.80\%$$

$$0.12\% < \pi_K < 0.50\%$$

Propulsion Systems, continued

Typical Grain Design Configurations

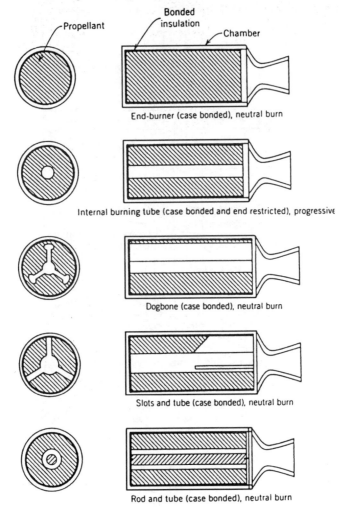

End-burner (case bonded), neutral burn

Internal burning tube (case bonded and end restricted), progressive

Dogbone (case bonded), neutral burn

Slots and tube (case bonded), neutral burn

Rod and tube (case bonded), neutral burn

Propulsion Systems, continued

Star (neutral)

Wagon Wheel
(neutral)

Multiperforated
(progressive-regressive)

Dendrite
(case bonded)

Classification of Propellant Grains According to Pressure-Time Characteristics

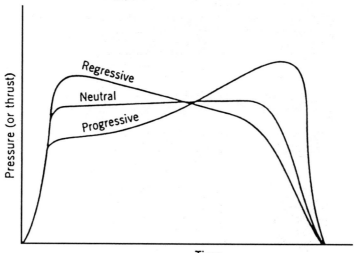

Propulsion Systems, continued

Theoretical Performance of Typical Solid Rocket Propellant Combinations

Oxidizer	Fuel	ρ,[a] g/cm^3	T_1, K	c^*,[b] m/s	$\mathfrak{M}_c$, kg mol	I_{sp},[b] s	k
Ammonium nitrate	11% binder and 7% additives	1.51	1282	1209	20.1	192	1.26
Ammonium perchlorate 78–66%	18% organic polymer binder and 4–20% aluminum	1.69	2816	1590	25.0	262	1.21
Ammonium perchlorate 84–68%	12% polymer binder and 4–20% aluminum	1.74	3371	1577	29.3	266	1.17

[a]ρ, average specific gravity of propellant.
[b]Conditions for I_{sp} and c^*: combustion chamber pressure, 1000 psia; nozzle exit pressure, 14.7 psia; optimum nozzle expansion ratio; frozen equilibrium.

Propulsion Systems, continued

Characteristics of Some Operational Solid Propellants

Propellant type[a]	I_{sp} range,[b] s	Flame temp, °F	Density, lb/in.^{3}	Metal content, wt%	Burning rate,[c] in./s	Pressure exponent, n	Hazard classification[d]	Stress, psi/Strain, % −60°F	Stress, psi/Strain, % +150°F	Processing method
DB	220–230	4100	0.058	0	0.45	0.30	1.1 or 1.3	4600/2	490/60	Extruded
DB/AP/Al	260–265	6500	0.065	20–21	0.78	0.40	1.3	2750/5	120/50	Extruded
DB/AP-HMX/Al	265–270	6700	0.065	20	0.55	0.49	1.1	2375/3	50/33	Solvent cast
PVC/AP/Al	260–265	5600	0.064	21	0.45	0.35	1.3	369/150	38/220	Cast or extruded
PS/AP/Al	240–250	5000	0.062	3	0.31	0.33	1.3	320/11	99/42	Cast
PU/AP/Al	260–265	5400–6000	0.064	16–20	0.27	0.15	1.3	1170/6	75/33	Cast
PBAN/AP/Al	260–263	5800	0.064	16	0.55	0.33	1.3	520/16	71/28	Cast
								(at −10°F)		
CTPB/AP/Al	260–265	5600–5800	0.064	15–17	0.45	0.40	1.3	325/26	88/75	Cast
HTPB/AP/Al	260–265	5600–5800	0.067	4–17	0.40	0.40	1.3	910/50	90/33	Cast
PBAA/AP/Al	260–265	5400/6000	0.064	14	0.32	0.35	1.3	500/13	41/31	Cast
AN/Polymer	180–190	2300	0.053	0	0.3	0.60	1.3	200/5	n/a	Cast

[a] Al, aluminum; AN, ammonium nitrate; AP, ammonium perchlorate; CTPB, carboxy-terminated polybutadiene; DB, double base; HMX, cyclotetramethylene tetranitramine; HTPB, hydroxy-terminated polybutadiene; PBAA, polybutadiene–acrylic acid polymer; PBAN, polybutadiene–acrylic acid acrylonitrile terpolymer; PS, polysulfide; PU, polyurethane; PVC, polyvinyl chloride.
[b] At 1000 psia expanding to 14.7 psia.
[c] At 1000 psia.

Attitude Control

Types of Attitude Control Systems

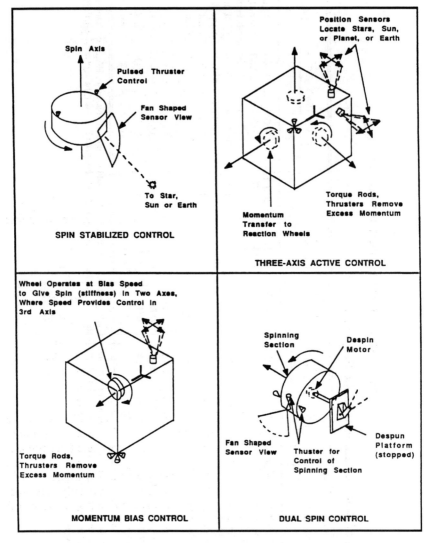

SPIN STABILIZED CONTROL

THREE-AXIS ACTIVE CONTROL

MOMENTUM BIAS CONTROL

DUAL SPIN CONTROL

Attitude Control, continued

Method	Typical accuracy	Remarks
Spin stabilized	0.1 deg	Passive, simple; single axis inertial; low cost
Gravity gradient	1–3 deg	Passive, simple; central body oriented; low cost
Jets	0.1 deg	Consumables; fast; high cost
Magnetic	1 deg	Near Earth; slow; low weight, low cost
Reaction wheels	0.01 deg	Internal torque, requires other momentum control; high power, cost

Coordinate Frames

The attitude of a satellite defines the relation between a reference coordinate frame and a satellite body-fixed coordinate frame.

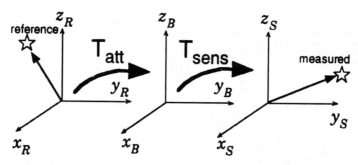

Measurements of reference vectors (e.g., sun, stars, local vertical) made in the sensor frame are used to compute the attitude matrix T_{att}. T_{att} is usually parameterized in Euler angles for analysis or Euler symmetric parameters (quaternions) for numerical computations.

The choice of an Euler angle sequence for the transformation T_{att} is dependent on the application, such as inertially stabilized, spin stabilized, or vertically stabilized. Proper choice of the Euler angle sequence will simplify the formulation. A typical Euler angle sequence (θ, ϕ, ψ) is shown in the following figure.

Attitude Control, continued

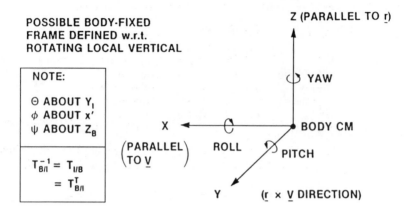

POSSIBLE BODY-FIXED FRAME DEFINED w.r.t. ROTATING LOCAL VERTICAL

NOTE:

Θ ABOUT Y_I
ϕ ABOUT x'
ψ ABOUT Z_B

$T_{B/I}^{-1} = T_{I/B}$
$\quad\;\; = T_{B/I}^{T}$

Z (PARALLEL TO $\underline{r}$)

YAW

X $\leftarrow$ BODY CM
(PARALLEL TO $\underline{V}$)

ROLL　PITCH

Y　($\underline{r} \times \underline{V}$ DIRECTION)

(PITCH, ROLL, YAW) = $(\Theta, \phi, \psi) \longrightarrow$ EULER ANGLES

TRANSFORMATION FROM BODY TO "INERTIAL" FRAME:

$$T_{B/I} = \begin{pmatrix} C_\psi & S_\psi & 0 \\ -S_\psi & C_\psi & 0 \\ 0 & 0 & 1 \end{pmatrix} \begin{pmatrix} 1 & 0 & 0 \\ 0 & C_\phi & S_\phi \\ 0 & -S_\phi & C_\phi \end{pmatrix} \begin{pmatrix} C_\Theta & 0 & -S_\Theta \\ 0 & 1 & 0 \\ S_\Theta & 0 & C_\Theta \end{pmatrix}$$
$$\quad\quad\quad\; \text{(YAW)} \quad\quad\quad\;\; \text{(ROLL)} \quad\quad\quad\; \text{(PITCH)}$$

Angular Momentum

$$\underline{H}_{total} = \sum_{i=1}^{n} \underline{r}_i \times m_i \underline{\dot{r}}_i \quad \begin{pmatrix} \text{collection of} \\ \text{point masses} \\ \text{at } \underline{r}_i \end{pmatrix}$$

The net torque from all forces is

$$\underline{T} = \sum_{i=1}^{n} \underline{\rho}_i \times \underline{F}_i = \sum_{i=1}^{n} \underline{\rho}_i \times m_i \underline{\ddot{r}}_i$$

$$\underline{T} = \frac{d\underline{H}}{dt} = (\underline{\dot{H}})_{\text{body frame}} + \underline{\omega} \times \underline{H}$$

$$\underline{\dot{H}} = \underline{T} - \underline{\omega} \times (\underline{I}\,\underline{\omega})$$

Attitude Control, continued

Euler Equations

In a body-fixed principal axis cm frame,

$$\dot{H}_1 = I_1\dot{\omega}_1 = T_1 + (I_2 - I_3)\omega_2\omega_3$$
$$\dot{H}_2 = I_2\dot{\omega}_2 = T_2 + (I_3 - I_1)\omega_3\omega_1$$
$$\dot{H}_3 = I_3\dot{\omega}_3 = T_3 + (I_1 - I_2)\omega_1\omega_2$$

No general solution exists. Particular solutions exist for simple torques. Computer simulation usually required.

An important special case is the torque-free motion of a (nearly) symmetric body spinning primarily about its symmetry axis, such as a spin-stabilized satellite. Thus,

$$\omega_x, \omega_y \ll \omega_z \triangleq \Omega$$
$$I_x \cong I_y$$

And

$$\dot{\omega}_x \cong -K_x\Omega\omega_y$$
$$\dot{\omega}_y \cong K_y\Omega\omega_x$$
$$\dot{\omega}_z \cong 0$$
$$\implies \omega_x = \omega_y = A\cos\omega_n t$$

where

$$K_x = \frac{I_z - I_y}{I_x}$$
$$K_y = \frac{I_z - I_x}{I_y}$$
$$\omega_n = \sqrt{K_x K_y}\,\Omega$$

Also,

$$I_z > I_x = I_y \quad \text{or} \quad I_z < I_x = I_y$$

yields ω_n real and sinusoidal motion at the nutation frequency ω_n.

Attitude Control, continued

Two cases exist:

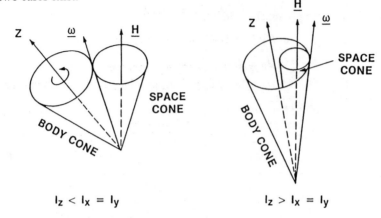

$$I_z < I_x = I_y \qquad\qquad\qquad I_z > I_x = I_y$$

The above figure shows the relationship between the nominal spin axis Z, the inertial angular velocity vector ω, and the angular momentum vector H. In the torque-free case, the angular momentum H is fixed in space, and the angular velocity vector ω rotates about H. For a disk-shaped body ($I_x = I_y < I_z$), the precession rate is faster than the spin rate. For a rod-shaped body ($I_x = I_y > I_z$), the precession rate is faster than the spin rate.

$$\Omega_P = \omega_Z \left[1 - \frac{I_T - I_S}{I_T} \right] = \omega_Z \left[\frac{I_S}{I_T} \right]$$

where

ω_Z = spin rate
Ω_P = precession rate

Spin Axis Stability

$$E_{total} = \frac{1}{2} \left(\sum_{i=1}^{n} m_i \right) R^2 + \frac{1}{2} \sum_{i=1}^{n} m_i \dot{p}_i^2$$

$$\underbrace{}_{E_{trans}} \qquad \underbrace{}_{E_{rot}}$$

For a rigid body, cm coordinates, with $\vec{\omega}$ resolved in body axis frame,

$$E_{rot} = \tfrac{1}{2}\underline{\omega} \cdot \underline{H} = \tfrac{1}{2}\omega^T I \underline{\omega}$$

The results above are valid only for a rigid body. When any flexibility exists, energy dissipation will occur, and $\underline{\omega}$ will lose its significance.

Attitude Control, continued

$$\underline{H} = \underline{\underline{I}}\,\underline{\omega} \Rightarrow \text{const}$$

$$E_{\text{rot}} = \tfrac{1}{2}\underline{\omega}^T\,\underline{\underline{I}}\,\underline{\omega} \Rightarrow \text{decreasing}$$

$\therefore$ Spin goes to maximum $\underline{\underline{I}}$, minimum $\underline{\omega}$.

Conclusion: Stable spin is possible only about the axis of maximum inertia. Classic example, Explorer 1.

Reference measurement methods

Reference	Typical accuracy	Remarks
Sun	1 min	Simple, reliable, low cost; not always visible
Earth	0.1 deg	Orbit dependent; usually requires scan; relatively expensive
Magnetic field	1 deg	Economical; orbit dependent, low altitude only; low accuracy
Stars	0.001 deg	Heavy, complex, expensive, accurate
Inertial space	0.01 deg/h	Rate only; good short-term ref; high weight, power, cost

The attitude control material appearing on pages 10-51–10-55 is from *Space Vehicle Design* by M. D. Griffin and J. R. French. Copyright © 1991, AIAA, Washington, DC. All rights reserved.

Power Systems

Power Supply Operating Regimes

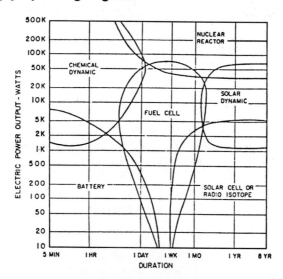

Power Systems, continued

Characteristics of Isotopes for Power Production

Isotope	Half-life	Compound	Compound power, W/g	Dose rate,[a] mR/h Bare	3-cm of uranium
Beta emitters					
Cobalt 60	5.27 yr	Metal	3.0[b]	3×10^8	6×10^6
Strontium 90	28 yr	$SrTiO_3$	0.2	6×10^6	1×10^4
Promethium 147[c]	2.67 yr	Pm_2O_3	0.3	1×10^5	1.0
Thulium 170	127 days	Tm_2O_3	1.75	4×10^6	50
Alpha emitters					
Polonium 210	138 days	Metal matrix	17.6	760	1.8
Plutonium 238	86 yr	PuO_2	0.35	5	0.03
Curium 242	163 days	Cm_2O_3 in metal matrix	15.5	280	2
Curium 244	18.4 yr	Cm_2O_3	2.5	600	32

[a]At 1 m from 5-thermal-kW source.   [b]200 Ci/g metal.   [c]Aged, 1 half-life.

Isotope Electrical Power vs Time

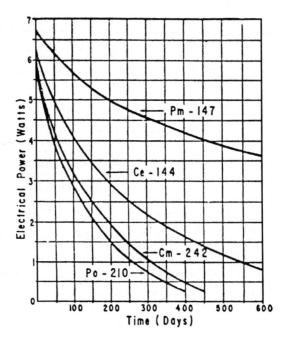

Power Systems, continued

Ni-Cd Battery Life

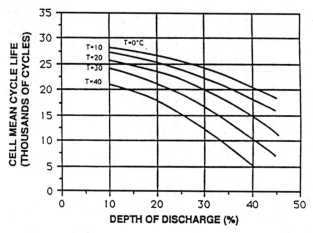

Ref.: *Spacecraft Systems Design Handbook* by W. J. Larson. Copyright © 1988, Kluwer Publishing. Reproduced with permission of Kluwer Publishing.

Characteristics of Different Storage Cells

Type of cells	Electrolyte	Nominal voltage/ cell, V	Energy density, Wh/kg	Temp, °C	Cycle life at different depth of discharge levels			Whether space qualified
					25%	50%	75%	
Ni-Cd	Diluted potassium hydroxide (KOH) solution	1.25	25–30	−10–40	21,000	3,000	800	Yes
Ni-H$_2$	KOH solution	1.30	50–80	−10–40	>15,000	>10,000	>4000	Yes[a]
Ag-Cd	KOH solution	1.10	60–70	0–40	3,500	750	100	Yes
Ag-Zn	KOH solution	1.50	120–130	10–40	2,000	400	75	Yes
Ag-H$_2$	KOH solution	1.15	80–100	10–40	>18,000	——	——	No
Pb-acid	Diluted sulfuric acid	2.10	30–35	10–40	1,000	700	250	——

[a]Ni-H$_2$ cells are employed onboard the Navigational Technology Satellite (NTS-2) and other geosynchronous satellites. However, these cells have not been used on any low-Earth orbit satellites.

Power Systems, continued

Expected Radiation Dose

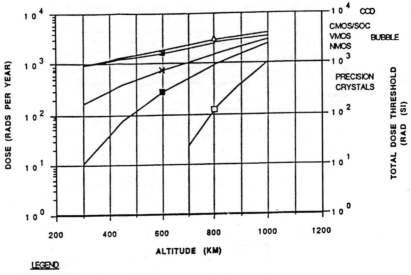

LEGEND

INCLINATION:

- ■ 30 deg
- ✳ 45 deg
- △ 60 deg
- ■ 90 deg
- ☐ 0 deg

100 mils Al
Spherical Shielding
Epocn = 198

Ref.: *Spacecraft Systems Design Handbook* by W. J. Larson. Copyright © 1988, Kluwer
Publishing. Reproduced with permission of Kluwer Publishing.

Power Systems, continued

Characteristics of a Typical Solar Cell Subjected to Successive Doses of 1 MeV Electrons

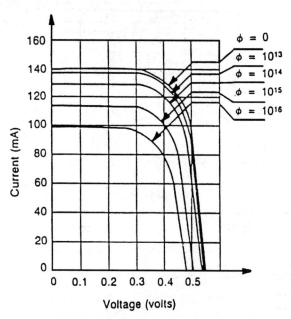

Ref.: *Spacecraft Systems Design Handbook* by W. J. Larson. Copyright © 1988, Kluwer Publishing. Reproduced with permission of Kluwer Publishing.

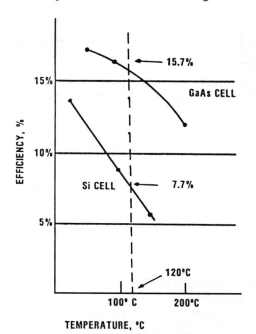

Power Systems, continued

Characteristics of a Typical Solar Cell for Various Temperatures[a]

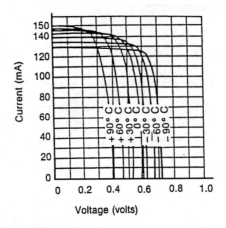

Eclipse Time vs Beta[a]

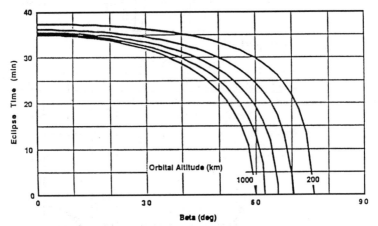

[a]Ref.: *Spacecraft Systems Design Handbook* by W. J. Larson. Copyright © 1988, Kluwer Publishing. Reproduced with permission of Kluwer Publishing.

$$T_e = (P/180) * \cos^{-1}\left[\left(1 - R_e^2/R^2\right)^{0.5} / \cos(\beta)\right]$$

$$P = 2 * \pi * [R^3/398{,}600.8]^{0.5}/60$$

where

T_e = eclipse time, min
P = orbital period, min
R_e = radius of Earth, 6378.135 km
R = distance from the center of Earth to the s/c, km
β = sun/orbit angle, deg

Power Systems, continued

Altitude Effects on Eclipses

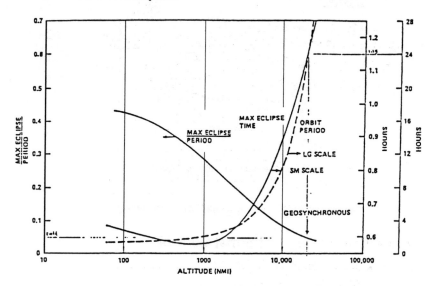

Solar Array Average Power vs Beta Angle

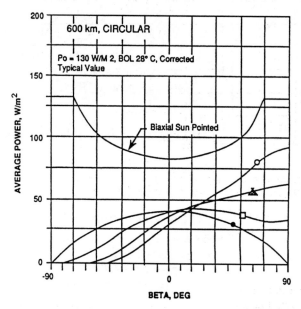

Ref.: *Spacecraft Systems Design Handbook* by W. J. Larson. Copyright © 1988, Kluwer Publishing. Reproduced with permission of Kluwer Publishing.

Power Systems, continued

LEGEND

- O FIXED, 0° CANT
- ☐ FIXED, 15° CANT
- ⬟ FIXED, 30° CANT
- o FIXED, 45° CANT

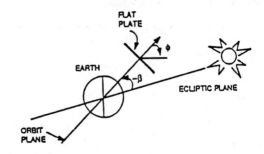

β = ANGLE BETWEEN THE ECLIPTIC PLANE AND THE ORBIT PLANE

φ = CANT ANGLE

NADIR POINTED SPACECRAFT WITH A FLAT, ROOFTOP SOLAR ARRAY

Thermal Control

Fundamentals of Heat Transfer

Conduction

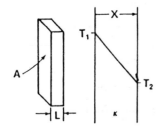

where

$$\dot{Q} = \text{power} = \frac{\text{energy}}{\text{time}} = W$$

κ = thermal conductivity (W/mK)
A = surface area, m^2
T = temperature, K

$$\dot{Q} = -\kappa A \frac{dT}{dx}$$ 1-D conduction equation (Fourier's Law)

$$\dot{q} \triangleq \frac{\dot{Q}}{A} = -\kappa \frac{dT}{dx}$$ 1-D heat flux equation

$$\underline{\dot{q}} = -\kappa \nabla \underline{T}$$ 3-D conduction equation

$$\nabla^2 T = \frac{\rho C}{\kappa} \frac{\partial T}{\partial t} - \frac{\dot{g}_{int}}{\kappa}$$ from 1st law thermodynamics

where

ρ = density, kg/m^3
C = specific heat, J/kgK
$\dot{g}_{int}$ = internal source, W/m^3

Thermal Control, continued

Cases

No source   ⇒ diffusion equation
Source but   ⇒ Poisson equation
  steady state
steady, no   ⇒ Laplace equation
  source

Convection

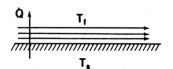

where
h = convection or film coefficient, W/m²K
$\dot{Q}$ = power, W
A = surface area, m²
ΔT = temperature difference for process, K

$\dot{Q} = hA\Delta T$ 1-D convection equation

$T_a > T_f,$

$\Delta T = T_s - T_f$ forced convection condition

Radiation

$$\dot{Q} = \epsilon A \sigma T^4 \qquad \text{Stefan–Boltzman law}$$

where

σ = Stefan–Boltzman constant 5.6696×10^{-8} W/m²K⁴
A = surface area, m²
ϵ = emissivity (blackbody $\overset{\Delta}{=} 1$)

$$E \overset{\Delta}{=} \frac{\dot{Q}}{A} = \text{hemispherical total emissive power, W/m}^2$$

$$E_{\lambda b} = \frac{2\pi h c^2}{\lambda^5 (e^{hc/\lambda kT} - 1)} = \frac{\text{Planck's law (vacuum)}}{\text{hemispherical spectral emissive power}}$$

where

$h = 6.626 \times 10^{-34}$ Js
$k = 1.381 \times 10^{-23}$ J/K
$c = 2.9979 \times 10^8$ m/s

Thermal Control, continued

Wien's displacement law—peak intensity of $E_{\lambda b}$:

$$(\lambda T)_{\text{MAXIMUM}} = \frac{1}{4.965114} \frac{hc}{k} = 2897.8 \; \mu \text{ m K}$$

Blackbody radiation fundamentals:
- Neither reflects nor transmits incident energy; perfect absorber at all wavelengths, angles
- Equivalent blackbody temperature sun $\cong$ 5780 K; Earth $\cong$ 290 K

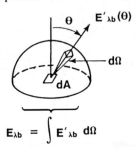

$$E_{\lambda b} = \int E'_{\lambda b} \; d\Omega$$

Lambertian surface:

$\qquad$ = Spectral directional radiant intensity, W/m^2 μm sr

$$i'_{\lambda b}(\lambda) = \frac{1}{\pi} E_{\lambda b}(\lambda):$$

$\qquad$ = Energy per unit time, per unit solid angle, per unit projected area $dA_{\perp}$, per unit wavelength
$\qquad$ = Directional spectral emissive power

$$E'_{\lambda b} = i'_{\lambda b} \cos \theta:$$

$\qquad$ = Energy per unit time, per unit wavelength, per unit solid angle, per unit area dA

Thermal Control, continued

Surface Properties for Passive Thermal Control[a]

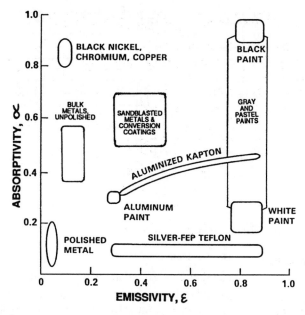

[a] Absorptivity values are at BOL (beginning of life).

Radiation Properties of Various Surfaces[a]

Surface description	Solar absorptivity alpha	Infrared emissivity epsilon
Al foil coated w/10 Si	0.52	0.12
Si solar cell 1 mm thick	0.94	0.32
Si solar cell 3 mil glass	0.93	0.84
Stainless steel, type 410	0.76	0.13
Flat black epoxy paint	0.95	0.80
White epoxy paint	0.25	0.88
2 mil Kapton w/vacuum dep Al	0.49	0.71
first surface mirror		
vacuum deposited gold	0.3	0.03
vacuum deposited Al	0.14	0.05
second surface mirror		
6 mil Teflon, vacuum dep Ag	0.09	0.75
6 mil Teflon, vacuum dep Al	0.14	0.75

[a] Absorptivity values are nominal (at BOL).

Ref.: *Spacecraft Systems Design Handbook* by W. J. Larson. Copyright © 1988, Kluwer Publishing. Reproduced with permission of Kluwer Publishing.

Thermal Control, continued

Equilibrium Temp vs Absorptivity to Emittance Ratio ($\beta = \alpha/\varepsilon$)[a]

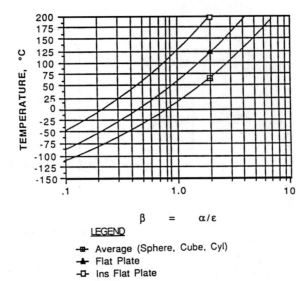

$$\beta \quad = \quad \alpha/\varepsilon$$

LEGEND

-▣- Average (Sphere, Cube, Cyl)
-▲- Flat Plate
-□- Ins Flat Plate

[a]Does not account for shadowing, albedo, Earthshine, or internal heat dissipation. It is appropriate for MLI covers, solar arrays dissipation, and passive structure.

Ref.: *Spacecraft Systems Design Handbook* by W. J. Larson. Copyright © 1988, Kluwer Publishing. Reproduced with permission of Kluwer Publishing.

Equations

$$T_{ES} = [(\beta * S * A_P/A_T * A_T)/(A_R * \sigma)]^{0.25}$$

$$T_{ESA} = [((\alpha - F_P * e) * A_F * S * A_P)/(\varepsilon F * A_F + \varepsilon B * A_B)]^{0.25}$$

where

β = absorbance/emissivity, α/ε
S = solar constant = 1356 w/m^2
A_P = projected area, m^2
A_T = total area, m^2
A_R = radiator area, m^2
A_B = back of solar array area, m^2
A_F = frontal area of solar array, m^2
σ = Boltzman constant = 5.97×10^{-8} w/m^2
α = absorbance
F_P = packing factor for solar array
e = solar cell efficiency
εF = emissivity of front of solar array
εB = emissivity of back of solar array

Thermal Control, continued

Assumptions

$A_F = A_B$
$F_P = 0.95$
$A_T = A_R$
Unblocked view to space
Isothermal spacecraft
No MLI

Inputs Required

β or temperature, °C

Spacecraft Systems Design and Engineering Thermal Control

ENERGY BALANCE:

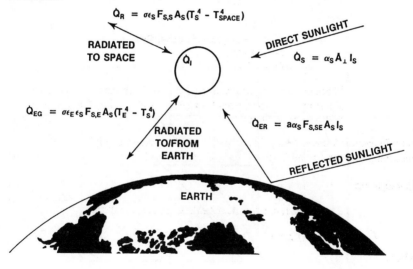

$$\dot{Q}_R = \sigma \epsilon_S F_{S,S} A_S (T_S^4 - T_{SPACE}^4)$$

RADIATED TO SPACE

DIRECT SUNLIGHT

$$\dot{Q}_S = \alpha_S A_\perp I_S$$

$$\dot{Q}_{EG} = \sigma \epsilon_E \epsilon_S F_{S,E} A_S (T_E^4 - T_S^4)$$

$$\dot{Q}_{ER} = a\alpha_S F_{S,SE} A_S I_S$$

RADIATED TO/FROM EARTH

REFLECTED SUNLIGHT

EARTH

Ref.: *Space Vehicle Design* by M. D. Griffin and J. R. French. Copyright © 1991, AIAA, Washington. DC. All rights reserved.

Thermal Control, continued

where

a = Earth albedo (0.07–0.85)
α_S = spacecraft absorptivity (solar)
$A_\perp$ = orbit-averaged spacecraft projected area perpendicular to sun
A_S = spacecraft surface area
ϵ_S = spacecraft emissivity (IR)
ϵ_E = Earth emissivity (IR) $\cong 1$
$F_{S,S}$ = view factor, spacecraft to space
$F_{S,E}$ = view factor, spacecraft to Earth
$F_{S,SE}$ = view factor, spacecraft to sunlit Earth
I_S = solar intensity, 1356 W/m^2
T_E = Earth blackbody temperature, 261 K

$$F_{S,S} + F_{S,E} = 1$$

If $\epsilon_E = 1$, $T_{\text{SPACE}} = 0$, and $F_{S,S} + F_{S,E} = 1$, solve to yield the average satellite temperature.

$$\sigma \epsilon_S A_S T_S^4 = \sigma \epsilon_S F_{S,E} A_S T_E^4 + \dot{Q}_S + \dot{Q}_{ER*} + \dot{Q}_1$$

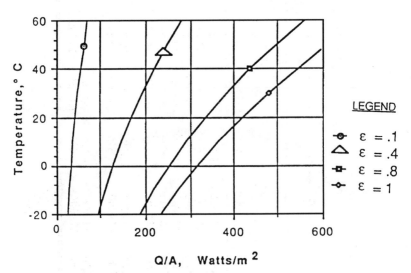

LEGEND

   $\epsilon = .1$
   $\epsilon = .4$
   $\epsilon = .8$
   $\epsilon = 1$

Q/A,   Watts/m^2

Ref.: *Spacecraft Systems Design Handbook* by W. J. Larson. Copyright © 1988, Kluwer Publishing. Reproduced with permission of Kluwer Publishing.

Fluid Systems for Heat Transfer

Fluid	Symbol	Molecular weight	Critical point			Properties Triple point			
			P, MPa	T, K	ρ, M/L	P, MPa	T, K	ρ, M/L Liquid	Vapor
Helium	He	4.0026	0.2275	5.2014	17.399	0.00496	2.172	36.5343	0.2904
Nitrogen	N_2	28.0134	3.4100	126.26	11.21	0.1246	63.15	30.977	0.2396
Oxygen	O_2	31.9988	5.0422	154.481	13.63	0.149 E−03	54.359	40.620	0.3275 E−03
Argon	Ar	39.948	4.8980	150.86	13.41	0.0689	83.8	35.40	0.1015 E−02
Methane	CH_4	16.0430	4.599	190.55	10.23	0.01174	90.68	28.1511	0.0157 E−03
Hydrogen (normal)	H_2	2.0159	1.3152	33.217	14.936				
Water	H_2O	18.0153	22.1123	647.383	17.74				
Carbon dioxide	CO_2	44.0100	7.375	304.217	10.64				
Hydrazine	N_2H_2	32.050	14.690	653.16					
Ammonia	NH_3	17.0306	11.422	406.161	13.835				

Communications

Gain vs Frequency

On both transmit and receive, for 100% efficiency,

$$G = \frac{4\pi}{\lambda^2} A \qquad \lambda = \frac{c}{f}$$

$$\therefore G = \frac{4\pi f^2}{c^2} A$$

Note that the higher the frequency, the greater the gain for fixed antenna size. And, for small beamwidth angles θ (deg),

$$G\theta^2 \cong 27,000 \deg^2$$

$$\therefore \theta \cong 164 \frac{\lambda}{\pi D} \deg \quad (\text{parabolic})$$

where

D	= diameter
f	= frequency
c	= speed of light
Beamwidth	= angle between -3 dB points (half-power points)
Gain	= antenna gain, on-axis, relative to ideal isotropic radiator (0 dB)

From energy conservation, ideally,

$$G\phi = 4\pi\, sr = 41,253 \deg^2$$

whereas in practice, for a typical parabolic dish,

$$G\phi = 2.6\pi\, sr - 27,000 \deg^2$$

is more likely. There will be a tradeoff between transmitter power and beam pointing.

Communications, continued

Antenna Pattern

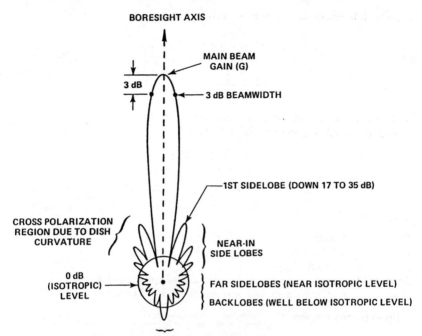

BORESIGHT AXIS

MAIN BEAM
GAIN (G)

3 dB

3 dB BEAMWIDTH

1ST SIDELOBE (DOWN 17 TO 35 dB)

CROSS POLARIZATION
REGION DUE TO DISH
CURVATURE

NEAR-IN
SIDE LOBES

0 dB
(ISOTROPIC)
LEVEL

FAR SIDELOBES (NEAR ISOTROPIC LEVEL)

BACKLOBES (WELL BELOW ISOTROPIC LEVEL)

REAR REGION AROUND AXIS (LEVEL MAY BE ELEVATED
ABOVE BACKLOBES BY 10 dB OR MORE)

Communications, continued

Typical Space Communications Antennas

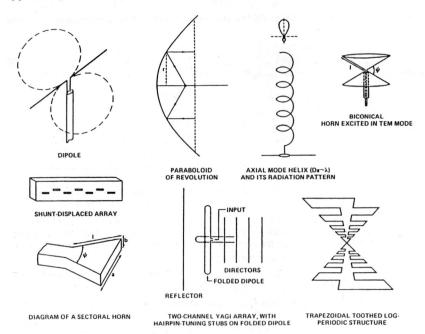

Gain and Effective Area of Several Antennas

Type of antenna	Gain	Effective area
Isotropic	1	$\dfrac{\lambda^2}{4\pi}$
Elementary dipole	1.5	$1.5\dfrac{\lambda^2}{4\pi}$
Halfwave dipole	1.64	$\dfrac{1.64\lambda^2}{4\pi}$
Horn (optimum)	$\dfrac{10A}{\lambda^2}$	$0.81A$
Parabolic reflector (or lens)	$6.2\text{–}7.5\dfrac{A}{\lambda^2}$	$0.5A\text{–}0.6A$
Broadside array (ideal)	$\dfrac{4\pi A^2}{\lambda^2}$	A

Communications, continued

Noise Sources

Sun	$T = 10^4 - 10^{10}$ K
	Communication is effectively impossible with the sun in the antenna fov
Moon	Reflected sunlight, but much weaker; communication usually unimpeded
Earth	Characteristic temperature 254 K
Galaxy	Negligible above 1 GHz
Sky	Characteristic temperature 30 K
Atmosphere	Noise radiated by O_2 and H_2O absorption and re-emission, typically less than 50 K below 20 GHz
Weather	Heavy fog, clouds, or heavy rain can outweigh other sources (except sun), especially above 10 GHz
Johnson noise	Due to resistance or attenuation in medium (atmosphere, wire, cable, etc.); proportional to temperature, resistance
Electronics noise	Receiving equipment (antenna, amplifiers) makes a significant contribution, typically 60 K cryogenically cooled equipment may be used for extremely low signal levels; pre-amplifier customarily mounted on antenna

Composite Link Noise Plot (Excluding Weather)

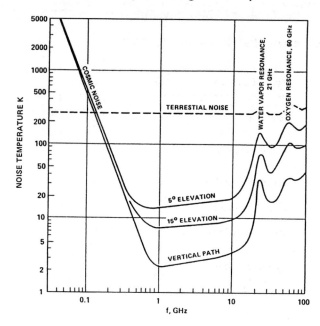

Communications, continued

Receiver Noise

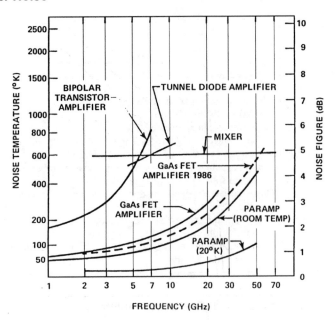

Total Absorption Loss (Excluding Weather)

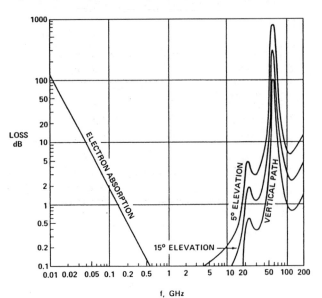

Communications, continued
Behavior of Analog and Digital Modulations

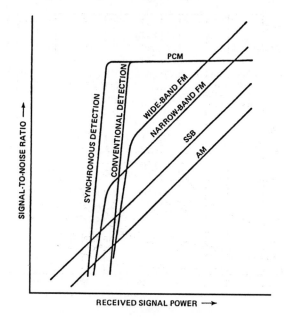

Structure Design and Test

Acceleration

Booster	Axial, g	Lateral, g
Atlas–Centaur	5.5	2
Titan II	3–10	2.5
Ariane	4.5	0.2
Delta	5.8–6.9	2.5
	5.7–12	
Long March	4.1	0.6
Shuttle	$-3.17 + 4.5$	$+ -1.5$
		$+ 4.5 - 2$

Structure Design and Test, continued

Low-Frequency Vibration

Vehicle	Frequency range	Acceleration
Atlas–Centaur	0–50 Hz	1.0 g axial
		0.7 g lateral
Ariane	5–100 Hz	1.0 g axial
	2–18 Hz	0.8 g lateral
	18–100 Hz	0.6 g lateral
Delta	5–6.2 Hz	0.5* amplitude axial
	6.2–100 Hz	1.0 g axial
	5–100 Hz	0.7 g lateral
Long March	5–8 Hz	3.12 mm axial
	8–100 Hz	0.8 g axial
	2–8 Hz	2.34 mm lateral
	8–100 Hz	0.6 g lateral

Random Vibration

Vehicle	Frequency range	Acceleration
Atlas–Centaur	20–80 Hz	+9 dB/oct
	80–200 Hz	0.03 g^2/Hz
	200–1500 Hz	−9 dB/Hz
		2.7 gms overall
Ariane	20–150 Hz	+6 dB/oct
	150–700 Hz	0.04 g^2/Hz
	700–2000 Hz	−3 dB/Hz
		7.3 gms overall
Long March	10–100 Hz	+3 dB/oct
	100–800 Hz	0.04 g^2/Hz
	800–2000 Hz	−12 dB/oct
		6.23 gms overall
Shuttle	20–100 Hz	+6 dB/oct
	100–250 Hz	0.015 g^2/Hz sill
		0.15 g^2/Hz keel
	250–2000 Hz	−6 dB/oct
Acoustics		137 dB to 142 dB

Structure Design and Test, continued

Load Definitions

Load	Specified acceleration
Limit load	Maximum expected acceleration
Yield load	Member suffers permanent deformation
Ultimate load	Structural member fails
Safety factor	Ratio of ultimate load to limit load
Margin of safety	Safety factor minus one = SF − 1

Load Factors

Use of Load Factors

Gross factors used in preliminary design
Sometimes multiplied by an uncertainty factor
Design LL = load factor * uncertainty factor
Typical uncertainty factor < 1.5
Gross load factors replaced by nodal accelerations after coupled modes analysis

Limit, Yield, and Ultimate

Expendable launch vehicles
    Ultimate = 1.25 × limit
    Yield = 1.0 × limit
    Sometimes yield = 1.1 × limit
For shuttle ultimate = 1.4 × limit
Pressure vessels are higher
    Typically qualification load = 1.2 × limit
    Yield = 1.2 × 1.1 × limit
    Ultimate = 1.5 × limit

Structure Design and Test, continued

Spacecraft Structure Design/Verification

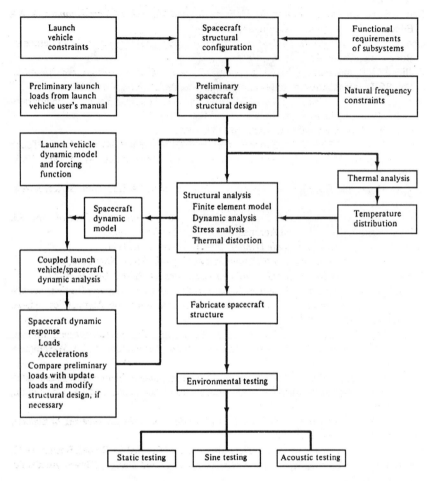

Ref.: *Design of Geosynchronous Spacecraft* by Agrawal, B. N. Copyright © 1986. Reprinted by permission of Prentice–Hall, Inc., Upper Saddle River, NJ.

References

Agrawal, B. N., *Design of Geosynchronous Spacecraft*, Prentice–Hall, Englewood Cliffs, NJ, 1986.

Battin, R. H., *An Introduction to the Mathematics and Methods of Astrodynamics*, AIAA Education Series, AIAA, Washington, DC, 1987.

Chetty, P. R. K., *Satellite Technology and its Applications*, TAB Professional and Reference Books, Blue Ridge Summit, PA, 1991.

Collicott, H. E., and Bauer, P. E. (eds.), *Spacecraft Thermal Control, Design, and Operation*, Vol. 86, Progress in Astronautics and Aeronautics, AIAA, New York, 1983.

Colwell, R. N. (ed.), *Manual of Remote Sensing*, 2nd Ed., American Society of Photogrammetry, Falls Church, VA, 1983.

Corliss, W., *Scientific Satellites*, NASA SP-133, 1969.

Garrett, H. B., and Pike, C. P., *Space Systems and Their Interactions with Earth's Space Environments*, Vol. 71, Progress in Astronautics and Aeronautics, AIAA, New York, 1980.

Griffin, M. D., and French, J. R., *Space Vehicle Design*, AIAA Education Series, AIAA, Washington, DC, 1991.

Horton, T. E. (ed.), *Spacecraft Radiative Transfer and Temperature Control*, Vol. 83, Progress in Astronautics and Aeronautics, AIAA, New York, 1982.

Hughes, P. C., *Spacecraft Attitude Dynamics*, Wiley, New York, 1986.

Kaplan, M. H., *Modern Spacecraft Dynamics and Control*, Wiley, New York, 1976.

Pocha, J. J., *Mission Design for Geostationary Satellites*, D. Reidel, Boston, 1987.

Shahrokhi, F., Greenberg, J. S., and Al-Saud, T. (eds.), *Space Commercialization: Launch Vehicles and Programs*, Vol. 126, Progress in Astronautics and Aeronautics, AIAA, Washington, DC, 1990.

Shahrokhi, F., Hazelrigg, G., and Bayuzick, R. (eds.), *Space Commercialization: Platforms and Processing*, Vol. 127, Progress in Astronautics and Aeronautics, AIAA, Washington, DC, 1990.

Shahrokhi, F., Jasentuliyana, N., and Tarabzouni, N. (eds.), *Space Commercialization: Satellite Technology*, Vol. 128, Progress in Astronautics and Aeronautics, AIAA, Washington, DC, 1990.

Sutton, G. P., *Rocket Propulsion Elements, An Introduction to the Engineering of Rockets*, 6th Ed., Wiley, New York, 1992.

Wertz, J. R. (ed.), *Spacecraft Attitude Determination and Control*, D. Reidel, Boston, 1978.

Wertz, J. R., and Larson, W. J., *Space Mission Analysis and Design*, Kluwer Academic, Boston, 1991.